The Emplacement of Silicic Domes and Lava Flows

Edited by

Jonathan H. Fink
Geology Department
Arizona State University
Tempe, Arizona 85287

SPECIAL PAPER
212

© 1987 The Geological Society of America, Inc.
All rights reserved.

Copyright is not claimed on any material prepared
by government employees within the scope of their
employment.

All materials subject to this copyright and included
in this volume may be photocopied for the noncommercial
purpose of scientific or educational advancement.

Published by The Geological Society of America, Inc.
3300 Penrose Place, P.O. Box 9140, Boulder, Colorado 80301

GSA Books Science Editor Campbell Craddock

Library of Congress Cataloging-in-Publication Data

The Emplacement of silicic domes and lava flows.

(Special paper ; 212)
"Outgrowth of a symposium, 'The emplacement of silicic
domes', held at the 1984 Geological Society of
America meeting in Reno, Nevada"—Pref.
Includes bibliographies.
1. Domes (Geology)—Congresses. 2. Lava—
Congresses. I. Fink, Jonathan H. II. Geological
Society of America. Meeting (1984 : Reno, Nev.)
III. Series: Special paper (Geological Society of
America) ; 212.
QE611.E47   1987   551.2'1   86-31807
ISBN 0-8137-2212-8

This book is dedicated to Don Allen,
Bob Compton, and Nina DeLange.

Cover photos:
Upper left: Santiaguito dacite dome complex, Santa Maria Volcano, Guatemala, viewed from the south. Photo taken February 1983 by Maurice Krafft.
Upper right: Mount St. Helens Dome, Washington, viewed from the south rim of the crater. Photo taken April 15, 1983, by Lyn Topinka, U.S. Department of Interior, U.S. Geological Survey, David A. Johnston, Cascades Volcano Observatory, Vancouver, Washington.
Lower left: South Sister Volcano, Oregon, viewed from south. Devil's Hill rhyodacite dome chain extends back from middle foreground toward South Sister. Photo by C. Dan Miller, U.S. Geological Survey, Cascades Volcano Observatory, Vancouver, Washington.
Lower right: Glass Mountain rhyolite flow, Medicine Lake Highland Volcano, California, viewed from northwest. Photo by Ron Greeley, Geology Department, Arizona State University, Tempe, Arizona.

Contents

Preface v

Growth of the lava dome at Mount St. Helens, Washington, (USA), 1981–1983 1
D. A. Swanson, D. Dzurisin, R. T. Holcomb, E. Y. Iwatsubo, W. W. Chadwick, Jr.,
T. J. Casadevall, J. W. Ewert, and C. C. Heliker

Volcanic activity at Santiaguito volcano, 1976–1984 17
William I. Rose

Eruptive histories of Lipari and Vulcano, Italy, during the past 22,000 years 29
Michael F. Sheridan, G. Frazzetta, and L. La Volpe

Holocene rhyodacite eruptions on the flanks of South Sister volcano, Oregon 35
William E. Scott

Tephra deposits associated with silicic domes and lava flows 55
Grant Heiken and Kenneth Wohletz

Origin of pumiceous and glassy textures in rhyolite flows and domes 77
Jonathan H. Fink and Curtis R. Manley

*Textural heterogeneities and vent area structures in the 600-year-old lavas of
the Inyo volcanic chain, eastern California* 89
Daniel E. Sampson

Types of mineralization related to fluorine-rich silicic lava flows and domes 103
Donald M. Burt and Michael F. Sheridan

An extensive, hot, vapor-charged rhyodacite flow, Baja California, Mexico 111
Brian P. Hausback

*Physical features of rhyolite lava flows in the Snake River Plain
volcanic province, southwestern Idaho* 119
Bill Bonnichsen and Daniel F. Kauffman

Preface

The blocky, inhospitable surfaces of silicic lava domes, along with their relatively small volumes and restricted global distribution, have until recently relegated them to positions of obscurity in the geologic literature. However, interest in the emplacement of silicic domes has increased rapidly in the last decade, stimulated by several factors. Foremost among these has been the continued, and in some cases predictable, extrusion of dacite lava in the crater of Mount St. Helens. Second, renewed deformation and seismicity in California's Long Valley caldera—where the last eruptions less than 600 years ago produced silicic domes and pyroclastic deposits—has demonstrated the need for more comprehensive models relating volcanic hazards to the variety of extrusive modes for silicic magma. The exploration for geothermal resources and volcanogenic ore deposits has encouraged studies of both the geometry of conduit systems beneath domes and their flow mechanisms once extruded. Finally, two scientific drilling programs in the Long Valley and Valles Calderas, sponsored by the U.S. Department of Energy, National Science Foundation, and U.S. Geological Survey, have provided the first interior views of very young silicic dome and conduit systems. Among the many studies spawned by these drill cores have been several concerning the role of volatiles in eruption and surface flow processes.

This Special Paper is an outgrowth of a symposium, "The emplacement of silicic domes," held at the 1984 Geological Society of America meeting in Reno, Nevada. The symposium emphasized the mechanical, rather than chemical aspects of dome emplacement, and this volume maintains that approach. Most of the central questions being addressed today about silicic domes were first posed a half century or longer ago. Many of these earliest studies are summarized in Williams' (1932) comprehensive monograph "The history and character of volcanic domes," and in Macdonald's (1970) book *Volcanoes*. The papers in this volume consider domes from several different perspectives, and offer insight into many of the key questions that still remain unanswered.

The AGI Glossary (Bates and Jackson, 1980) defines volcanic domes as "steep-sided, rounded extrusions of highly viscous lava squeezed out from a volcano, and forming dome-shaped or bulbous masses of congealed lava above and around the volcanic vent." Here we restrict our attention to domes of silicic composition (dacite to rhyolite) because of their closer association with geothermal activity, ore deposition, and the hazards of explosive volcanism. Most of the papers are concerned with silicic "domes" rather than "flows," although in many cases a sharp division between the two is not apparent. In kinematic terms, domes may be considered to be extrusions in which most of the movement is upward, whereas flows, or coulees, are emplaced with a significant amount of lateral advance away from the vent.

Other problematic distinctions concern the modes of extrusion for silicic lavas. Two styles of dome growth that have commonly been inferred (and recently observed at Mount St. Helens) are "endogenous," in which new lava is added to the interior; and "exogenous," in which all fresh lava is extruded on the dome's surface. At the other end of the volumetric spectrum, the largest silicic lava flows are easily confused with certain pyroclastic deposits. Complete welding and remobilization of some extensive ash flows can result in rocks that appear virtually identical to lavas, both in hand specimen and under the microscope. These so-called "rheomorphic" tuffs are the subject of considerable current research.

Formulation of models of dome emplacement must start with observations and measurements of active extrusions. Direct observation of growing domes has always been both hazardous and expensive. Furthermore, the lifetime of an individual dome-producing vent may extend for decades or even centuries, making most long-term studies logistically overwhelming. The first two papers in our collection present accounts and interpretations of two of the most thoroughly studied, currently active volcanic domes of the Western Hemisphere: one in the crater of Mount St. Helens, Washington, and the other near the summit of Santa Maria Volcano in Guatemala.

At Mount St. Helens, members of the U.S. Geological

Survey have carefully monitored the episodic growth of the composite dacite dome since the first lobe appeared in June of 1980. The dome's first two years were characterized by alternating periods of exogenous extrusion and endogenous inflation, each cycle lasting a few weeks (Moore and others, 1981). During the following three-year period, the pattern became less regular; continuous extrusion lasted for periods of as long as six months. The first paper presented here, by Swanson and others, chronicles the growth of the dome during the critical period from 1981 to 1983, and presents some suggestions about how changes in extrusion rate and eruptive style may reflect changes taking place in the underlying magma chamber and in the internal structure of the dome itself.

Cyclic dome growth like that exhibited by Mount St. Helens may continue for periods of years, decades, or centuries. The paper by Rose presents observations of the Santiaguito dacite dome complex at Santa Maria Volcano in Guatemala, which has been active since 1923. Although it is less frequently monitored than Mount St. Helens, Santiaguito offers a continuous record of extrusive activity covering more than half a century. The recognition that periods of quiet dome emplacement have alternated with more hazardous explosive activity has allowed Rose to speculate on the timing and magnitude of future eruptions from Santiaguito.

As longer cycles of dome emplacement are recognized, steady-state processes of magma production can be defined. The paper by Sheridan, Frazzetta, and La Volpe describes such extended cycles of dome emplacement, based on their mapping and dating of rhyolite domes and coeval pyroclastic deposits (ranging up to 22,000 yr B.P. in age) from the Italian islands of Lipari and Vulcano. Citing the spatial distribution, volumes, and ages of the domes and pyroclastic deposits, they were able to model and compare the production rates of the individual vents, and to distinguish different eruptive mechanisms. As at Santiaguito, the long-term record of eruptive activity has allowed Sheridan and his colleagues to make long-range assessments of volcanic hazards related to the silicic vents on Lipari and Vulcano. However, the lack of detailed deformation data for individual eruptions prevents them from making the types of short-term eruption forecasts now possible at Mount St. Helens.

Several of the papers in this volume concern relationships between dome growth and pyroclastic activity, because of the direct bearing these have on hazards assessment, and because of the insights they provide into processes within magma chambers and conduits. In the fourth paper, Scott describes the late Holocene eruptive history of South Sister Volcano in the Cascade Range of Oregon. This activity included emplacement of a well-preserved series of rhyodacite domes and flows, and associated pyroclastic flows, air-fall deposits, and lahars. Through careful mapping, Scott established stratigraphic and temporal relationships among the domes and other eruptive products. These relations, along with structural and chemical evidence, permit estimates of the size and location of the magma chamber, and constrain the en echelon geometry of the dikes that carried magma from this chamber to the surface. The proposed emplacement of domes from dikes is similar to that inferred for many sets of domes (Fink and Pollard, 1983) and is in apparent contrast to the central pipelike conduit feeding lava to the growing dome at Mount St. Helens.

Silicic domes are usually found with tephra deposits that formed before, during, and after lava extrusion. In the fifth paper, Heiken and Wohletz present a model for explosive activity accompanying dome growth, and use scanning electron microscopic, sedimentologic, and theoretical studies to arrive at a set of criteria for determining when in an eruptive cycle a given tephra deposit formed. Such distinctions are crucial to both the successful prediction of hazards associated with silicic dome emplacement, and the accurate determination of the eruptive history of a given vent. They illustrate their model with examples from several different dome-producing eruptions.

Much of the explosive eruptive activity that precedes dome growth results from the exsolution of volatiles before magma reaches the surface. Evidence for the distribution of magmatic volatiles both before and during extrusion is found in the textures and volatile contents seen in silicic domes. Fink and Manley present microscopic and macroscopic observations of pumiceous, glassy, and crystalline textures from the surfaces of numerous sets of domes and flows, and from drill cores of the Inyo Scientific Drilling Program. To explain the observed distribution of textures, flow banding, and water content, they use a model in which volatiles migrate upward, through microcracks, from crystallizing interior zones to the base of a flow's pumiceous carapace. They conclude that, although explosive hazards generally decrease during the emplacement of silicic domes, migration and concentration of volatiles within larger flows can eventually result in a tendency for explosive activity that increases with distance from a vent.

The 550-year-old Inyo domes in eastern California are among the most thoroughly studied silicic extrusions in the world. In the paper by Sampson, new mapping of the distribution of textures on the three youngest Inyo Domes is integrated with chemical, mineralogical, stratigraphic, and structural data to arrive at a new emplacement model. Building on several earlier studies of the domes and their explosive counterparts, Sampson presents evidence that two magma types (finely and coarsely porphyritic) were stratified within the feeder dike prior to eruption. He also shows structural relations in the vent areas of the domes which suggest that their near-surface conduits were circular or pipelike, rather than elongate, in cross section. This paper clearly demonstrates how structural and chemical studies of domes can be synthesized to derive new interpretations of emplacement mechanisms.

A different kind of synthesis is presented in the paper by Burt and Sheridan, who discuss factors influencing the mineralization of fluorine-rich silicic domes and lava flows. Three different types of economically important deposits are considered, along with the emplacement conditions believed necessary for their formation. Beryllium-rich deposits, such as those at Spor

Mountain, Utah, are found in tuffs immediately underlying silicic domes. Mapping and mass balance calculations suggest that these tuffs are mineralized in response to devitrification of the overlying lavas. Fumarolic, "Mexican type" tin deposits are shown to be formed in the brecciated carapaces of silicic domes. Finally, Climax or porphyry-type molybdenum and tungsten deposits are related to devolatilization of highly fractionated magmas beneath intrusive domes and small plutons.

Depending upon topography, chemistry, and eruption rates, some silicic domes evolve into flows that are capable of traveling long distances before coming to rest. The last two papers describe extensive rhyolite flows whose problematic identities typify the difficulties involved in distinguishing large rhyolite flows from remobilized ash-flow tuffs. In Hausback's study of the Providencia rhyolite from southern Baja California, he first demonstrates that the numerous fault blocks in which the lava is exposed represent the same flow. He then presents structural and textural evidence supporting the interpretation of Providencia as a lava flow rather than a remobilized tuff. Finally, he cites mineralogic data indicating that the lava had unusually high eruption temperatures and volatile contents, both of which could have contributed to the anomalously low viscosity required to form a 27-km-long flow.

In the final paper of the collection, Bonnichsen and Kauffman describe a series of controversial rhyolite lavas in southwest Idaho that have previously been interpreted as remobilized tuffs. In the course of their arguments about the origin of these flows, the authors present a detailed catalogue of the types of structures typically observed in rhyolite lavas, a model for their development, and a list of features used to distinguish lava flows from pyroclastic flows. In a final cautionary note they warn that some voluminous, high-temperature tuffs may become so totally welded that they develop the same types of structures observed in lava flows, and lose all evidence of their pyroclastic origins. They suggest that future experimental and theoretical studies be directed toward developing petrologic criteria for distinguishing flows of explosive and effusive origin.

These ten papers provide only a sampling of the research being conducted on silicic domes. Most of the studies begin with field observations and mapped stratigraphic relations, and attempt to explain either the petrologic consequences, hazards, or economic aspects of dome and flow emplacement. The overall shapes of some domes reflect the rheology of their magma, as illustrated by field, experimental, and theoretical studies (e.g., Huppert and others, 1982). Several of the papers in this volume relating lava textures to volatile contents rely on earlier experimental studies such as those by Lofgren (1971), and water content measurements such as those by Eichelberger and Westrich (1981) and by Rutherford and others (1985).

Domes form at various intervals during the lifetimes of silicic volcanoes and alternate with explosive activity; a major emphasis of dome studies is on the understanding of these cycles. Silicic domes are the surface expressions of complex conduit systems that carry magma upward from chambers at depth and control interactions with meteoric water. The geometry of these conduits provides important information, not only about subsurface distributions of magma, but also about the prevalent stress conditions controlling eruption and emplacement (e.g., Bacon, 1985).

Advances in the modeling of dome emplacement will require the acquisition of geophysical and geologic data from many more volcanoes. Such efforts will be assisted by the development of airborne and satellite imaging systems capable of the high resolution necessary to create topographic maps. Such maps can then be used to trace and define the deformation patterns of actively growing domes. Deformation networks that can be read remotely after initial deployment will also provide valuable data. Another area that needs further exploration concerns the physical properties of silicic magmas. In particular, the temperature and volatile content dependence of such rheologic properties as viscosity, yield strength, and tensile strength need to be better established. Understanding the interaction between local and regional stresses in controlling the distribution of domes (e.g., Bacon and others, 1980) will help in the interpretation of older sets of domes.

It is hoped that the studies described in this Special Paper will stimulate further investigations of physical processes controlling the shallow emplacement of silicic magma, just as earlier work by Howell Williams and his predecessors inspired many of the contributors to this volume.

REFERENCES CITED

Bacon, C. R., 1985, Implications of silicic vent patterns for the presence of large crustal magma chambers: Journal of Geophysical Research, v. 90, p. 11243–11252.

Bacon, C. R., Duffield, W. A., and Nakamura, K., 1980, Distribution of Quaternary rhyolite domes of the Coso Range, California; Implication for extent of the geothermal anomaly: Journal of Geophysical Research, v. 85, p. 2425–2433.

Bates, R. L., and Jackson, J. A., eds., 1980, Glossary of geology, 2nd edition: Falls Church, Virginia, American Geological Institute, 751 p.

Eichelberger, J. C., and Westrich, H., 1981, Magmatic volatiles in explosive rhyolitic eruptions: Geophysical Research Letters, v. 8, p. 757–760.

Fink, J. H., and Pollard, D. D., 1983, Structural geologic evidence for dikes beneath silicic domes, Medicine Lake Highland Volcano, California: Geology, v. 11, p. 458–461.

Huppert, H. E., Shephard, J. B., Sigurdsson, H., and Sparks, R.S.J., 1982, On lava dome growth with application to the 1979 lava extrusion of the Soufriere of St. Vincent: Journal of Volcanology and Geothermal Research, v. 14, p. 199–222.

Lofgren, G., 1971, Experimentally produced devitrification textures in natural rhyolitic glass: Geological Society of America Bulletin, v. 82, p. 111–124.

Macdonald, G. A., 1970, Volcanoes: Englewood Cliffs, New Jersey, Prentice Hall, 510 p.

Moore, J. G., Lipman, P. W., Swanson, D. A., and Alpha, T. R., 1981, Growth of lava domes in the crater, June 1980-January 1981, in Lipman, P. W., and Mullineaux, D. R., eds., The 1980 eruptions of Mount St. Helens: U.S. Geological Survey Professional Paper 1250, p. 541–547.

Rutherford, M. J., Sigurdsson, H., Carey, S., and Davis, A., 1985, The May 18, 1980, eruption of Mount St. Helens; 1. Melt composition and experimental phase equilibria: Journal of Geophysical Research, v. 90, p. 2929–2947.

Williams, H., 1932, The history and character of volcanic domes: University of California Publications in Geological Sciences, v. 21, p. 241–263.

MANUSCRIPT ACCEPTED BY THE SOCIETY MAY 5, 1986

Growth of the lava dome at Mount St. Helens, Washington, (USA), 1981-1983

D. A. Swanson
D. Dzurisin
R. T. Holcomb
E. Y. Iwatsubo
W. W. Chadwick, Jr.
T. J. Casadevall
J. W. Ewert
C. C. Heliker
U.S. Geological Survey
Cascades Volcano Observatory
5400 MacArthur Boulevard
Vancouver, Washington 98661

ABSTRACT

Nine dominantly nonexplosive episodes of dome growth at Mount St. Helens during 1981–83 added about 40×10^6 m^3 of dacitic lava to the active composite dome in the volcano's 1980 crater. Endogenous and exogenous growth, the latter mostly in the form of stubby lava flows that accumulated on the dome, combined to build an edifice 880 m long, 830 m wide, and 224 m high by December 1983; the total volume (1980–1983) was about 44×10^6 m^3. Every 1–5 months during 1981–82, periods of increasing seismicity and ground deformation lasting 1–3 weeks culminated in extrusions lasting a few days. Endogenous growth became increasingly important during this interval, and in February 1983, the style of activity changed from episodic to essentially continuous endogenous and exogenous growth.

In March 1982 and February 1983, extrusions were preceded by lateral explosions that triggered snow avalanches from the crater wall and mudflows down the volcano's north flank. Collapse of part of the north face of the dome during rapid endogenous growth in April 1982 produced a hot rock avalanche and small mudflow. Ejections of gas and comminuted dome rock from the top of the dome were frequent throughout 1981–83, and averaged several per day in 1983.

As the dome continues to grow, its capacity to accommodate newly supplied magma without rupturing, and hence the ratio of endogenous to exogenous growth, will probably continue to increase. Endogenous growth will be especially favored when magma supply is continuous and slow; extrusive activity will be favored if the short-term supply rate becomes significantly higher than the current rate of about 0.9×10^6 m^3/month. Future explosive activity is possible and perhaps even likely in view of Mount St. Helens' history and the histories of similar contemporary domes at other volcanoes, but at this time we can foresee no change from the pattern of mostly nonexplosive dome growth that characterized activity during 1981–83.

INTRODUCTION

The explosive activity at Mount St. Helens during 1980 was only the beginning of an eruptive process that continues. The 1980 activity was described by Christiansen and Peterson (1981) and Moore and others (1981); this paper updates that chronology through the end of 1983. We also compare the current Mount St. Helens lava dome to those at Mount Lamington (Papua New Guinea), Bezymianny (Kamchatka, USSR), and Santa Maria Santiaguito (Guatemala), and comment on possible future developments at Mount St. Helens. Swanson and others (1983, 1985) gave a description of the methods used to monitor and predict episodes of dome growth during 1981–82; additional references to specific monitoring results are provided in the narrative below. Waitt and others (1983, Fig. 2) provided a location map showing the 1980 crater, the lava dome, and the area affected by the 1981–83 activity. Detailed topographic maps are being prepared for a more thorough analysis of dome growth; sketch maps and volume estimates in this paper are subject to revision. This paper is primarily a chronology; interpretations of ongoing activity continue to evolve and will be subjects of future papers.

CHRONOLOGY

The great rockslide avalanche, lateral blast, and Plinian eruption at Mount St. Helens on May 18, 1980 were followed by five smaller explosive magmatic eruptions between May 25 and October 18 of that year (Christiansen and Peterson, 1981). Dacite domes were emplaced on the floor of the May 18 crater during the waning stages of three of those events, but the June and August 1980 domes were mostly destroyed by explosive eruptions in July and October, 1980, respectively. The October 1980 dome survived and forms the oldest part of the currently active composite dome. In December 1980, the first in a series of extrusions added material to the north and south flanks of the October 1980 dome, and formed a composite dome having a volume of about 4×10^6 m^3 (Moore and others, 1981).

Generally nonexplosive dome growth continued during 1981–83; intrusion and extrusion combined to increase the volume of the dome to about 44×10^6 m^3 by December 1983 (Table 1). Growth of the dome during that time is illustrated schematically in Figures 1 and 2, and dimensional information is provided in Table 1. For brevity, the figures and table are not repeatedly referenced, but the reader should refer to them while reading the narrative.

The 1981–83 dacite is chemically and petrographically similar to the 1980 dacite (Moore and others, 1981). Both are highly microphyric mafic dacite containing about 62.5 wt% SiO_2 and phenocrysts of plagioclase, hypersthene, and hornblende in descending order of abundance (Cashman and Taggart, 1983). Inclusions are dominantly of gabbro, and form 2–5 vol.% of the dacite (Heliker, 1984).

The dome has grown chiefly by the extrusion of stubby lava flows, commonly termed lobes, onto its flanks and adjacent crater floor. However, the dome has also enlarged internally by endogenous growth. Evidence of endogenous growth comes from observations of morphologic changes without associated extrusion and from instrumental measurements, principally by electronic distance meters, of deformation of the dome. Commonly minor endogenous growth precedes extrusion, and its detection forms an important means of anticipating future extrusion. As the dome has enlarged, the ratio of endogenous to exogenous (extrusive) growth has increased. The following narrative describes the complex interplay between these two different but related processes.

February 1981

The extrusion of December 1980 (Moore and others, 1981) was followed by a similar event during February 5–7, 1981. The February extrusion was heralded in mid-January by accelerating deformation of the crater floor, including movement along a preexisting thrust fault tangential to the dome and opening of cracks radial to the dome (Chadwick and others, 1983). On February 3, workers in the crater felt numerous earthquakes and observed new glowing cracks on the dome and new impact craters, radial cracks, and thrust faults on the crater floor. We don't know how much if any endogenous growth occurred as magma rose into the dome, but suspect that it was minor compared to the volume eventually extruded.

Following a continued increase in shallow seismicity on February 4 (Malone and others, 1983), a new lobe was first seen atop the dome at midday on February 5. The onset of extrusion was not observed, but a decline in the number of earthquakes shortly after dawn on February 5 may indicate the time when magma gained easy access to the surface of the dome. The new lobe welled out of a preexisting collapse pit in the October 1980 lobe and eventually buried all but small parts of that lobe. The north December 1980 lobe was disrupted and mostly overridden, and a small segment of the south December 1980 lobe was also covered. An extensive talus apron developed on the dome's western flank and spread onto the adjacent crater floor.

Geodetic measurements showed that the dome grew higher by about 30 m during February 5–6, before the spreading rate of the new lobe's eastern front slowed from 1.0 m/hr on February 7 to 0.5 m/hr on the afternoon of February 8. The summit began to subside when extrusion slowed, and by February 8 the rate of subsidence had reached 0.2 m/hr. Most extrusion was probably over by this time, but the new lobe and the rest of the dome continued to settle and spread slowly until the next extrusion; the top of the new lobe subsided almost 8 m between February 7 and 21.

April 1981

Another extrusion during April 10–12, 1981 added a lobe to the north-northwest flank of the dome. Precursory ground

TABLE 1. DOME GROWTH AT MOUNT ST. HELENS, 1980-83

Onset of Extrusion	Duration (days)	Erupted Volume (10^6 m^3)	Dome Volume[1] (10^6 m^3)	Dome Dimensions[2] (l/w/h in m)
6/13/80[3]	7	5	5	365/365/45
8/08/80[3]	2	1	1	120/120/60
10/18/80	1	1	1	300/300/34
12/27/80	7	3	4	440/320/84
2/05/81	3	4	8	500/420/119
4/10/81	3	3.5	11.5	680/440/113
6/18/81	2	4.5	16	680/500/163
9/06/81	5	4.5	20.5	680/560/164
10/30/81	4	2	22.5	700/580/182
3/19/82	24	5[4]	26.5	780/600/204
5/14/82	5	2.5	29	780/600/191
8/18/82	6	1	30	840/700/205
2/07/83[5]	328+	14.5	43.5	880/830/224
Total		51.5[6]	43.5[6]	

[1] Cumulative volumes shown at end of each extrusive episode, excluding talus that mantles crater floor at base of dome (probably less than 10% of volume of dome).

[2] Dimensions at end of each extrusive episode, including talus and displacements owing to endogenous growth: l, length (N/S); w, width (E/W); h, height above base of October 1980 lobe (now buried) based on vertical-angle measurements from fixed points.

[3] Destroyed by subsequent explosive eruptions in 1980; volumes poorly known.

[4] Includes 0.8 x 10^6 m^3 (dense rock equivalent) for March 19 pumice deposited beyond dome and 0.2 x 10^6 m^3 for April 4 rock-avalanche.

[5] Extrusion preceded by a lateral explosion and rockfall on February 2. Extrusion may have begun on February 5 and was confirmed on February 7. Endogenous growth during March-April 1983, with possible extrusion in March; combined endogenous and exogenous growth at other times.

[6] Erupted volume exceeds dome volume, because June and August 1980 domes (5 x 10^6 m^3 and 1 x 10^6 m^3, respectively) were largely destroyed by subsequent explosions; significant volumes of erupted material were deposited beyond dome by March 19, 1982 explosion (0.8 x 10^6 m^3) and by April 4, 1982 rock-avalanche (0.2 x 10^6 m^3); and a 1 x 10^6 m^3 notch was created in dome and then refilled in February 1983.

deformation began in late March, about 3 weeks before extrusion. Movement along thrust faults and radial cracks on the crater floor accelerated during March 21-28 (Swanson and others, 1983, Figs. 2 and 6). Leveling surveys that month revealed substantial uplift of the crater floor centered on the dome. Similar deformation patterns preceded the December 1980 and February 1981 extrusions; we attribute them to intrusion of magma beneath and eventually into the dome from a shallow source (Chadwick and others, 1983). Shallow earthquakes increased slightly in late March and then sharply on April 9.

At 0821 PST (8 hr earlier than GMT) on April 10, a tephra-laden plume penetrated clouds over the volcano and eventually rose 2.5 km above the crater floor. This was followed at 0824 by a gas plume containing little or no ash and at 0920 by a second ashy plume. Deteriorating weather prevented further observations, but shallow seismicity remained at a high level throughout the day before decreasing on April 11.

Helicopter views on the morning of April 12 revealed a new lobe on the north flank of the dome. A spine 8-10 m high and 3-4 m wide at its base towered above the new lobe but collapsed two days later. The new lobe, like all others on the dome, had a scoriaceous carapace and a relatively dense core, as revealed by deep cracks (Fig. 3). Its surface was undulatory in plan view; it resembled glacier ogives and suggested several overlapping surges perhaps caused by changes in extrusion rate. The April 1981 lobe eventually covered the northern part of the dome and extended its margin about 180 m (including talus) to the north-northwest. In addition, the crater floor was severely broken by new radial cracks and thrust faults.

Trilateration shortly after extrusion stopped showed that points on the crater floor and near the northern base of the pre-April dome had moved radially outward as much as 1.5 m since early March. These displacements probably resulted from swelling around the subsurface conduit before and during the extrusion. Projection of the displacement vectors back toward the dome suggests that the April extrusion was fed from the same

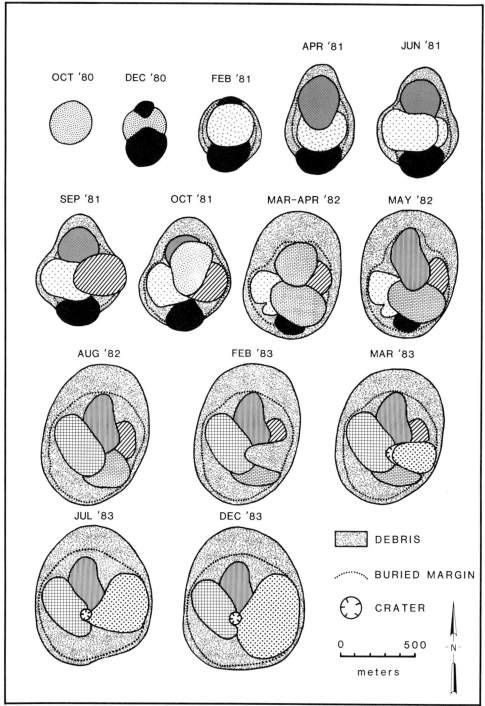

Figure 1. Sequential planimetric sketch maps from vertical air photos of composite lava dome at Mount St. Helens during 1981-83. Areas labeled debris include parts of dome and talus highly disrupted from rockfalls originating on the dome, excluding April 1982 rock-avalanche that extended well beyond north base of dome. Dotted line marks inferred contact between talus-mantled dome and talus-mantled crater floor; latter is much more extensive on north side of dome owing to northward slope of crater floor. For October 1980 until August 1982, dome is portrayed at end of each discrete growth episode; during 1983, four stages of continuous growth are shown. In March-April 1982 sketch, southern new lobe was extruded in March, followed by northern lobe in April. February 1983 sketch shows configuration of dome after February 2-4 explosions, before new lobe began to grow in resulting notch. Several features mentioned in text, including February 1983 spine, March 1983 mound, and April 1983 spines are not shown at this scale for clarity. Patterns other than that for October 1980 (upper left) same as those used in Figure 2; scale is approximate.

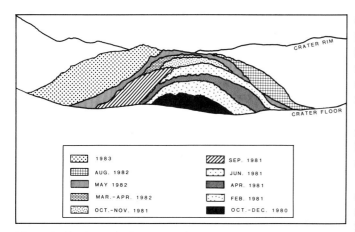

Figure 2. Tracings of photographs from fixed point on crater floor illustrates dome growth at Mount St. Helens during 1981–83. Products of each extrusive episode include talus that accumulated during that episode, unlike portrayal in Figure 1. Southward view from about 1 km north of center of dome. Scale varies owing to perspective view; true east-west basal diameter of dome in December 1983 was about 830 m. Photos and tracings by L. J. Topinka.

conduit as the February extrusion, located directly below the buried collapse pit on the October 1980 lobe. Some endogenous growth undoubtedly occurred before magma reached the surface of the dome, but subsequent exogenous growth dominated volumetrically.

June 1981

Rates of ground deformation and sulfur-dioxide emission increased in late May 1981, and on May 29 an electronic tiltmeter near the northeast base of the dome began to record tilting of the crater floor outward from the dome (Chadwick and others, 1983; Casadevall and others, 1981; Dzurisin and others, 1983). The rate of ground deformation accelerated through June 17, when seismicity and SO_2 emission rate both increased. A reversal in tilt direction and a change from impulsive earthquakes to emergent, low-frequency events suggest that the anticipated extrusion began the following night (Malone and others, 1983).

A new lobe was first seen in poor weather at midday on June 19; better observations on the 20th indicated only minor changes since the previous day, suggesting that the lobe had been mostly or wholly in place when it was first sighted. The June lobe was extruded from a collapse pit atop the February lobe; from there it descended the southwest flank of the dome to the crater floor (Fig. 4). Rockfalls from the east edge of the new lobe eventually created an overhanging cliff that exposed relatively dense dacite capped by a scoriaceous carapace 1–2 m thick.

September 1981

Following an interval of quiescence after the June extrusion, deformation of the crater floor resumed in early July and accelerated through early August. Vigorous gas and tephra emissions from the top of the dome (see cover of *Science,* v. 221, no. 4618, September 30, 1983), which had earlier occurred only once or twice daily, increased sharply on August 5 to produce a persistent ash-laden plume rising 1 km above the vent. The frequency of these degassing events declined to one every few days by August 13, then increased again on August 25. Sulfur-dioxide emission rates increased sixfold during August 18–24, but returned to a background level of about 60 t/day on August 26; no significant changes in seismicity occurred during this interval.

Deformation of the crater floor continued to accelerate until early on September 6, when the temporal frequency of shallow earthquakes increased sharply. A rare coincidence of daylight and favorable weather provided good views during the onset of extrusion. Fracturing and visible uplift of the northeast flank of the dome caused frequent rockfalls that generated large dust plumes, raised by convection to 1 km above the crater floor. By midmorning on September 6, the crater seismic record was dominated by rockfall signals; discrete earthquakes had become relatively rare.

Rockfall activity peaked at midday, before extrusion was observed. By late afternoon, a fracture high on the dome's east flank had become a small, smooth-surfaced spreading center (similar to that in Fig. 4) from which a viscous lava flow had begun to spread. Earthquakes and rockfall activity waned considerably throughout the afternoon as rising magma gained easy access to the surface of the dome. Airborne observers that night reported numerous incandescent areas on the surface of the new lobe and witnessed glowing rockfalls from its toe as the lobe crept downslope.

Shortly after dawn on September 7, the new extrusion appeared as a stubby lobe perched atop the east flank of the dome (Fig. 5). Rockfalls had removed much of the eastern carapace of the June 1981 lobe, and the new lobe was descending the resulting talus slope at an average rate of 1.6 m/hr. It slowed progressively to 0.3 m/hr during September 8–10 and to 0.1 m/hr during September 10–12. Extrusion stopped shortly thereafter and the dome began to stabilize.

October–November 1981

Thrust and tear faults on the crater floor continued moving at unusually high rates following the September extrusion (Chadwick and others, 1983, Figs. 2 and 3); this suggested that additional dome growth might occur soon. New hairline cracks appeared on the southwest crater floor during October 9–12, and others continued to form and widen until the anticipated extrusion began. Observers in the crater felt many small earthquakes beginning in mid-October, while deformation rates of the crater floor accelerated. The daily number of shallow earthquakes beneath the dome increased gradually from October 19 until the morning of October 30. Shallow earthquakes then decreased rapidly, and surficial rockfall and degassing events increasingly dominated the seismic record (see Malone and others [1983] for

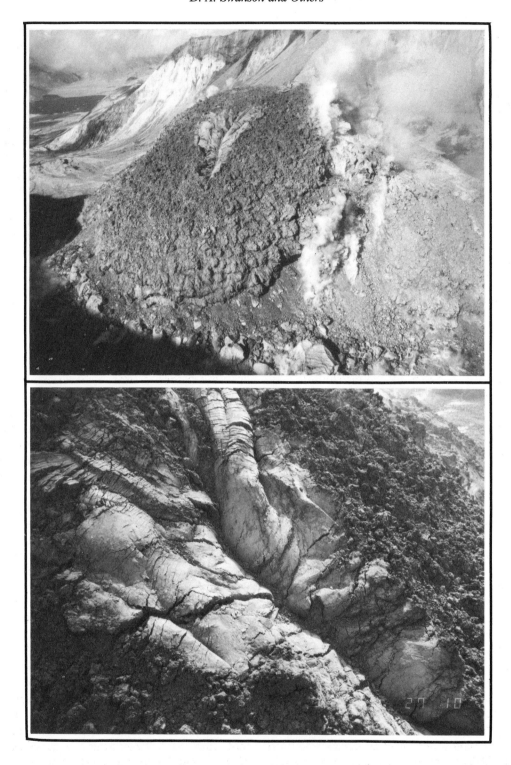

Figure 3. Top: View northeastward on June 26, 1981, of smooth spreading center atop June 1981 lobe. Dome approx. 160 m high. Part of crater floor and northeast crater wall visible between dome and Spirit Lake in background at upper left. Bottom: Detailed view of same extrusion center on August 27, 1981; width of view about 30 m. Photo by Terry Leighley.

distinctions between the seismic signatures of earthquakes, rockfalls, and degassing events).

Extrusion probably began on October 30, but poor weather delayed direct observations until the next morning. A new lobe had issued from the top of the dome, and the April 1981 and June 1981 lobes had become highly fractured and disrupted by endogenous growth. At noon October 31, the northern front of the flow advanced down the flank of the dome at 4 m/hr; parts of the flow behind the front moved slightly faster, about 4.5 m/hr. The spreading rate dropped to about 3 m/hr by late afternoon and to 1 m/hr by the following day; spreading had nearly ceased by November 4.

Electronic distance and triangulation measurements from the crater floor to targets on the dome showed that intrusion of magma had caused the west side of the dome to move laterally as much as 11.8 m before and during extrusion. This was the first direct measurement of endogenous growth associated with extrusion; similar, though lesser, endogenous growth probably had accompanied the earlier extrusions.

A small, laterally directed hot explosion occurred during bad weather early in the October–November 1981 event, presumably from a vent on the north flank of the dome. It produced a deposit of new ash up to 4 cm thick that extended more than 1.5 km north of the dome and formed a thin discontinuous layer south and east of the dome. The ash was hot enough to melt snow and partially melt thin sheets of plastic 300 m north of the dome. Wood and paint, however, were unaffected. Bent metal posts and damaged survey targets indicate that the explosion was mostly directed northward. Similar effects had been noted during earlier extrusions, but their cause was less clear.

March-April 1982

The most explosive event at Mount St. Helens since October 1980 occurred on March 19, 1982. Shallow seismicity and deformation of the crater floor and dome increased above background level in mid-February and distinctly accelerated in early March. The first in a series of relatively deep (greater than 5 km) earthquakes beneath the volcano occurred on February 8. Numerous other earthquakes 5–10 km deep that had magnitudes less than 1 occurred during the next month (Weaver and others, 1983).

Deep seismicity declined markedly on March 12, but shallow seismicity and deformation rates of the crater floor and dome increased for another week. The occurrence of relatively deep earthquakes, unusually high-energy release from shallow earthquakes, and a short burst of volcanic tremor during the afternoon of March 19 were recognized as possible precursors to renewed explosive activity.

At 1932 PST March 19, a lateral explosion from the south side of the dome dislodged most of the snow from a 120 degree sector of the 500-m-high south crater wall. Waitt and others (1983) inferred that the resulting dislodged material—snow, newly erupted pumice, and lithic blocks—flowed around the

Figure 4. Large crack atop May 1982 dome lobe, showing typical scoriaceous carapace and dense interior. Width of view about 30 m. Photo by Lyn Topinka.

dome, sped northward through the breach, and cascaded down the north flank of the volcano to Spirit Lake and the North Fork Toutle River. Waitt and others (1983) estimated the flow's volume as 10^6–10^7 m^3, and also inferred that a hydraulically ponded, pumice- and ash-clogged lake as deep as 15 m formed between the dome and south crater wall as a result of the rapid snowmelt. A debris-laden flood drained the lake, coursed around the dome and descended the volcano's north flank as a viscous slurry of rock debris and water (Fig. 6); it had a peak discharge of at least 13,800 m^3/s (Waitt and others, 1983). Part of the mudflow reached Spirit Lake, but most was channeled down the North Toutle River where it evolved into a hyperconcentrated water flow before reaching the Cowlitz and Columbia rivers.

Immediately following the initial explosion, a tephra-bearing plume rose 14 km above the dome. This spectacular preface was followed by extrusion of a new lobe in the southeast sector of the dome during March 20–24. Extrusion slowed on March 22–23, but rates of displacement, especially of the steep northern flank of the dome, accelerated again as endogenous growth continued. Bad weather prevented observations of the crater from March 26 until April 4. Seismicity increased during the afternoon of April 4, and in mid-afternoon aerial observers saw a large rockfall from the north flank of the dome. Seismicity intensified during the early evening until an explosion at 2052 PST generated a plume that rose 6.5 km above the dome. The associated seismic event consisted of at least 3 pulses and persisted as a continuous signal for about 20 min. This unusually large, prolonged signal probably originated from a series of rock-

Figure 5. Sequential photographs looking southwestward from east crater floor showing growth of September 1981 lobe (arrows) on dome's east flank. Dome about 160 m high on September 10. Photos by Terry Leighley.

falls and avalanches during failure of part of the dome's north flank. Two more explosions at 0035 and 0039 on April 5 produced a plume that rose almost 9 km above the vent. Seismicity temporarily declined after the third explosion, but low-level harmonic tremor was recorded from 0226 until 0435 and vigorous degassing persisted until mid-afternoon.

The rock-avalanche deposit covered about 8×10^4 m^2 and had a volume of approximately 2×10^5 m^3. The deposit consisted of many avalanches emplaced in rapid succession, but most of the volume was included in the two largest avalanches. The largest, which probably occurred at 2052 on April 4, formed a lobate deposit 450 m long and as much as 5 m thick near its margins. The avalanche material was hot and melted snow and thereby generated a flood down the volcano's north flank that evolved into a mudflow before entering the North Toutle River valley. The April 4 mudflow was smaller than that of March 19 and did not reach Spirit Lake.

Accelerated spreading of the north flank of the dome in late March suggests that failure of the flank on April 4–5 was caused by its oversteepening during rapid endogenous growth. This interpretation is supported by the appearance of a new dacite lobe on April 6. The source of this lobe was beneath the north edge of the March lobe, directly above the main feeder conduit inferred from previous geodetic measurements. The new lobe crept down the north flank of the dome at a maximum rate of 4 m/hr on April 6; it slowed to 1.9 m/hr on April 7, 0.6 m/hr on April 8, and 0.3 m/hr on April 9. The net advance of the intact lobe during this interval was considerably less than that implied by these short-term rates, because rockfalls removed material almost continuously from the toe of the flow.

Figure 6. Aerial photograph of March 19, 1982 mudflow; view to south from above Harry's Ridge along southwest shore of Spirit Lake. Mudflow originated in southern part of crater, divided around lava dome (here obscured by clouds and shadow), then merged into single flow several hundred metres north of dome. After descending north flank of volcano, mudflow divided again. Small flow turned northeast and entered Spirit Lake (bottom left), but most went westward down North Fork Toutle River (bottom right).

May 1982

Both deformation of the crater floor and shallow seismicity increased again in early May 1982, and culminated on May 14 with another extrusion onto the dome. Following a week of bad weather that prevented access to the crater, observers on May 11 discovered hundreds of small ground cracks radial to the dome, as well as many small buckles and a few small thrust faults tangential to the dome. All of these structures were confined to the western crater floor and had formed since the explosion and mudflow of March 19, 1982, which had erased a similar set of features formed during earlier extrusions (Chadwick and others, 1983).

Distance and angle measurements from the crater floor to the surface of the dome confirmed that endogenous growth was taking place in early May. Targets on the dome moved slightly downward and several metres outward before extrusion began, as magma entered the dome and began to spread laterally within it. The rate of spreading accelerated as magma approached the surface of the dome. Endogenous growth was nonuniform and concentrated in the western sector of the dome, but the eventual extrusion came from the top of the dome and flowed primarily northward.

A weak, continuous tremorlike signal began shortly after midnight on May 13–14 and fluctuated in intensity for the next 6 hr. The appearance of the dome had changed considerably by dawn, and geodetic measurements revealed that the jumbled surface of the April 1982 lobe was moving laterally as much as 0.2 m/hr.

New lava had reached the surface of the dome by the morning of May 15, when the blocky surface of the new lobe was measured as advancing northward by as much as 3 m/hr. Upward growth of the new lobe directly above the feeder conduit was noted later that day. When first measured, the vertical growth rate was about 1 m/hr, but it gradually slowed during the afternoon. By May 19, lateral spreading had slowed to a few centimeters per hour, and the surface of the new lobe had started to subside; net subsidence during May 15–21 was approximately 20 m. During the same interval, the surface morphology of the new lobe changed from blocky to typically scoriaceous; degassing may have played a role, but the process was not observed and remains a puzzle.

Extrusion ceased on about May 20, but the dome continued to spread slowly; the surface of the new lobe subsided at a decreasing rate thereafter. Slow sagging and spreading of the dome occurred between each of the 1981–83 extrusions for which data are available; we presume that this occurs in response to gravitational stresses acting on the plastic interior of the dome, heated during intrusions and loaded by new material added to the surface of the dome.

August 1982

Deformation rates of the dome and crater floor briefly accelerated in mid-July, and crater seismicity increased at the end of the month before declining in early August. Deformation rates remained high but did not accelerate further until about August 8, when both seismicity and rates of dome spreading clearly began to increase and endogenous growth apparently began in earnest. Between August 10 and 18, cumulative uplift of more than 70 cm occurred on the southwest crater floor adjacent to the dome, and a new thrust fault developed on the floor about 100 m from the base of the dome; its frontal scarp eventually grew to a height of 3 m.

Deformation, seismicity, and sulfur-dioxide emission rates increased sharply on August 18; new material first appeared atop the dome shortly before noon, when a large ovoid block rose above the west side of the May 1982 lobe. Vertical-angle measurements to the top of the block recorded uplift of 6 cm/min (3.6 m/hr) for roughly 90 min, when the rate slowed to 2–2.5 cm/min (1.2–1.4 m/hr). The high point of the new lobe rose 0.4–1.0 m/hr on August 19, and then subsided during August 20–23 at a gradually decreasing rate. The most rapid lateral spreading of the active lobe measured was 17 m/hr during a short interval on August 19.

On August 18, the new lobe was light gray with a smooth ropy surface; during the next few days this changed to a rough scoriaceous surface typical of earlier lobes. Extrusion slowed and stopped during August 19–23.

1983

In 1983, the style of growth changed from episodic to continuous for reasons that are still unclear. Crater seismicity and sulfur-dioxide emission rates increased temporarily in January 1983; movement of the north flank of the dome accelerated slightly, but movement of the west flank did not. Other flanks of the dome could not be monitored because of poor weather during January and February. The following sequence of events was documented by sporadic observations and visits to the crater. On January 30 and 31, minor explosions were recorded seismically and ashy plumes were detected by control-tower radar at Portland International Airport (75 km south-southwest of the volcano). A sudden increase in the number and size of shallow earthquakes on February 1 was followed by additional explosions on February 2, 3, and 4.

The largest explosion, on February 2, generated an ashy plume to a height of 4 km above the dome; surface winds distributed fine airfall mostly northwestward. The event included a small lateral explosion that sprayed rocks and ash over the east-northeast crater wall, triggered snow avalanches, melted snow on the crater floor, and generated a small mudflow that barely reached Spirit Lake (Fig. 7). In addition, part of the east side of the dome collapsed into a rock avalanche. Stratigraphic relations indicate that this collapse preceded, and perhaps caused, the lateral explosion. The February 2 collapse and explosion produced an east-facing notch near the top of the dome, about 60–80 m deep at its head, 200–250 m long, and 80–100 m wide. The floor of the notch was relatively flat and contained a crater 10 m deep rimmed by coarse ejecta and talus. This notch was the site of subsequent smaller explosions and rockfalls on February 3 and 4.

Aerial observations on February 5 suggested that the notch had enlarged considerably; a smooth-sided block about 10 m × 20 m × 10 m high within the notch may have been a new extrusion. An unusual pattern of incandescence in the notch possibly caused by a small growing lobe was photographed that night, but a new lobe was not confirmed until February 7. By then, new lava had covered the floor of the notch and flowed onto the rockfall deposit of February 2–4. A pie-shaped wedge of the March 1982 lobe along the south side of the notch, source of numerous rockfalls as early as February 5, had been visibly uplifted and tilted eastward.

When next observed, on February 10, the new lobe had enlarged by about 30%, and a smooth-walled crease interpreted as a spreading center extended along the long axis of the lobe. Like some previous examples, this center was later disrupted by differential flow.

As the February 1983 lobe grew, it nearly filled the notch that had developed during February 2–4. However, the head of the notch was not filled; its western wall, cut into the eastern part of the August 1982 lobe, remained exposed and the space between this wall and the western flank of the growing lobe comprised a deep, craterlike pit. This pit was never entirely filled, and subsequent small bursts of gas and tephra enlarged it to produce a craterlike depression rimmed by ejecta.

Measurements on February 23 showed that the new lobe had advanced about 23 m eastward since February 11; occasional rockfalls from the margin of the lobe suggested that slow movement was continuing. By the next day a spine of viscous lava projected from the surface of the lobe, which continued its slow advance. The spine had grown considerably by February 28, but the new lobe had apparently ceased to flow. On March 2 the spine stood 60 m above its base and about 19 m above the top of the dome (Fig. 8). By March 9 the spine had crumbled to 4 rooted blocks surrounded by a pile of blocky rubble.

Extrusion of the February 1983 lobe and spine had essentially ended by March 1, but crater seismicity and spreading of the dome continued above background levels. The highest rate of sulfur-dioxide emission (360 t/day) since August 1981 was measured on March 10, and minor but continuing morphologic changes on the east half of the dome, as well as sounds of grinding and falling rocks heard while walking on the dome, indicated that endogenous growth continued throughout March. The most notable feature to develop during this interval was a mound

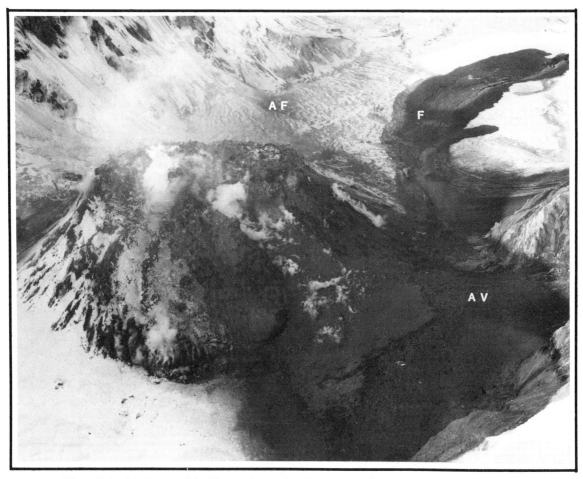

Figure 7. Aerial photograph looking northwest showing products of rock avalanche and lateral explosion from eastern sector of dome on February 2, 1983 (AV: rock avalanche from dome; AF: airfall; F: mudflow deposits). Explosion and avalanche created notch in dome just left of center (obscured by steam; see text) that was partly filled by extrusion of new lobe starting about February 7. Dome about 850 m wide. Photo by R. B. Waitt.

between the spine and the headwall of the February notch. This mound, about 30–40 m high and 100 m in basal diameter, formed an eastern rim for the residual crater-like pit at the head of the February notch. The mound probably resulted from upwelling of viscous lava from within the dome.

The mound grew a few meters higher and became the highest point on the dome by mid-April. Numerous slabs and small spines also grew from an area of jumbled topography east of the mound. By April 21, one of these had developed into a significant spine that became the highest point on the dome in late April. At the same time, large rockfalls from the active east flank were more frequent, and the rate of spreading of the dome increased. This activity culminated with extrusion of new lava from a broad area high on the northeast flank of the dome beginning between April 30 and May 4.

Slow extrusion and endogenous dome growth continued through June, accompanied by frequent small degassing bursts that sent plumes as high as 2 km above the top of the dome. These plumes issued from the large pit on top of the dome, directly above the feeding conduit as interpreted from geodetic data. Only ash and fine lapilli reached the crater floor during most of the bursts, but blocks as large as 30 cm in diameter peppered the top of the dome. Ejected material was relatively dense and was probably derived from within the dome or the upper part of the active feeding conduit.

Geodetic measurements of dome spreading during May–June 1983, together with visual observations, revealed that both endogenous and exogenous growth continued at nearly constant rates. The active lobe on the northeast flank of the dome, which was moving as much as 30 m/day when first observed in early May, continued to move 4–5 m/day through the end of June (although its net advance was much less owing to frequent rockfalls from its front). In addition, continued outward movement of 1–2 m/day of the northeast flank of the dome below the active lobe resulted in cumulative displacement of at least 50 m from mid-April to late June. This activity was accompanied by fre-

Figure 8. February 1983 spine viewed from southwest on March 2, 1983. Active east flank of dome warm enough to remain free of snow; elsewhere on dome, only hot cracks and steaming areas are free of snow. Spine is about 60 m high; part of northeast crater wall visible in upper right corner. Photo by Lyn Topinka.

quent ashy gas emissions (typically several per day) from several craters and fractures atop the dome. The largest crater had evolved from the unfilled part of the February notch, widened by rockfalls to a width of about 100 m and filled by debris to a depth of 30 m by the end of 1983.

Rates of endogenous and exogenous growth fluctuated slightly but remained high through September 1983, when a small "piggyback" lava flow oozed slowly onto the surface of the active lobe. The dark scoriaceous surface of the new outflow contrasted with the rest of the active lobe, which had not developed a scoriaceous carapace but instead was mantled by a jumble of dense gray blocks. The dark outflow probably began to spread in late August; by late September it was moving downslope at 1.4 m/day, roughly twice as fast as the underlying active lobe. What prompted the extrusion of the dark outflow is not clear. The chemical composition of the outflow is indistinguishable from that of all other extrusions during 1981-83; the main difference is textural. Perhaps differential spreading away from an area below the main vent, as suggested by triangulation data, created a weak zone through which lava oozed onto the surface of the active lobe. More rapid extrusion and degassing relative to the rest of the active lobe may have caused the scoriaceous carapace.

Extrusion of the dark outflow probably stopped sometime in September 1983, but extrusion on the dome's northeast flank and geodetically determined endogenous growth of the dome's east and southeast sectors continued through the end of the year. Advance of the active lobe on the northeast flank slowed as deformation rates of the southeast quarter of the dome accelerated greatly in early October, when a spine and ridge of new lava welled from the top of the dome. The southeast sector of the dome continued to deform rapidly through October. Deformation of this sector then slowed for the rest of the year, but in mid-December increased deformation, incandescence, and cracking became evident on the north and northeast flanks of the dome.

In summary, activity in 1983 was characterized by virtually continuous endogenous and exogenous growth that caused numerous rockfalls from the active east and southeast sectors of the dome. Endogenous growth, measured geodetically and recognized visually by disruption of surface features, played a major role in the growth. That lava which was extruded formed sluggish lobes less distinct than those of 1981–82. By December 31, 1983, the growing dome was approximately 880 m long, 830 m wide, 224 m high, and had an estimated volume of 44×10^6 m^3 (Figs. 9 and 10).

COMPARISON WITH OTHER DOMES WORLDWIDE

The Mount St. Helens dome can be placed in context by comparing it to well-known, long-lived, contemporary domes at Mount Lamington (Taylor, 1956, 1958), Bezymianny (Gorshkov, 1959; Kirsanov, 1979; Bogoyavlenskaya and Kirsanov, 1981; Seleznev and others, 1984), and Santiaguito (Rose, 1972, 1973). For a general summary, see Newhall and Melson (1983).

After a large explosive eruption at Mount Lamington on January 21, 1951, a dome grew rapidly in the summit crater to a volume of 1 km^3, accompanied by small explosions and pyroclastic flows. Taylor (1956) noted that the initial growth rate was remarkably rapid; during the first week, the dome attained a height of more than 300 m (reported as 1000 ft) above the crater floor! The rapidly growing dome collapsed on March 5, 1951, and generated lithic pyroclastic flows that removed about two-thirds of the dome's volume. The dome quickly rebuilt itself and grew rapidly through the middle of 1952, when activity started to decline; the dome stood approximately 550 m (1800 ft) above the crater floor in January 1953, 2 yr after the paroxysmal eruption.

Activity at Bezymianny volcano since 1955 has been similar in many respects to that at Mount St. Helens since 1980. After at least two centuries with no recorded eruptions, Bezymianny began a series of explosions in October 1955 that culminated on March 30, 1956 with a catastrophic lateral blast that formed an elongate crater 1.5 km wide and 0.7 km deep in the former summit area. More than 400 km^2 east of the volcano were devastated by the blast. Growth of the Novyy dome on the crater floor started in April 1956 and was accompanied by sporadic explosions that partly or wholly destroyed the young dome. Extrusive activity has dominated since 1965; the still-growing dome had a volume of about 1.6 km^3 at the end of 1982.

Historical activity at Santa Maria volcano in Guatemala commenced in October 1902, five months after smaller but more notorious eruptions at Mt. Pelee, Martinique, and Soufriere, St. Vincent. At Santa Maria, more than 5 km^3 of pyroclastic material was ejected during six explosive eruptions that year, most of it during the initial and strongest eruptions on October 24–26 (Rose, 1973). Following a 20-yr lull, growth of the Santiaguito dome commenced on the crater floor in June 1922, accompanied by sporadic ash ejections, pyroclastic flows, and lava flows. Activity, essentially continuous since 1922, had produced a composite dacite dome with a volume of about 0.8 km^3 by the end of 1982.

Mount St. Helens is currently the smallest of the four domes (Table 2) and has grown at the slowest rate, if we consider only the first 3 yr of growth for each dome (Fig. 11). Growth rates at

Figure 9. Comparative aerial views from north of 1980 crater on August 8, 1980 (top) and January 9, 1984 (bottom). East-west rim diameter of crater about 2.2 km. In bottom photo, note size of 1980–83 dome and deep gully incised into crater floor by surface runoff and groundwater sapping (lower left). 1980 photo by Terry Leighley; 1984 photo by Lyn Topinka.

Santiaguito and Bezymianny were virtually the same for the first 3 yr (about 5 times the rate for Mount St. Helens), and the Bezymianny dome has continued to grow at virtually the same rate for more than 27 yr. Growth of the Santiaguito dome slowed considerably after about 3 yr, and its average growth rate for the past 61 yr matches the Mount St. Helens rate for the past 3 yr; the current growth rate at Santiaguito is about half that at Mount St. Helens. The Mount Lamington dome grew more than 10 times faster than any of the other domes but stopped much sooner.

A rough inverse correlation exists between the average growth rate and silica content of these domes. The Mount Lamington and Bezymianny domes grew fastest, and their products are the most mafic. The Santiaguito and Mount St. Helens domes contain about 5% more silica than the others, and they have

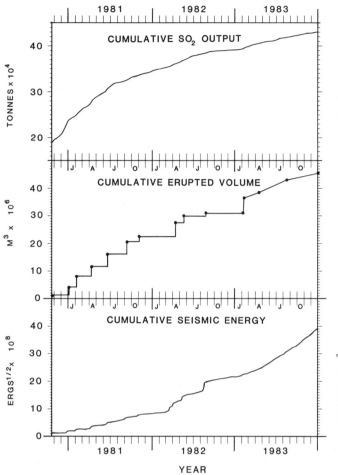
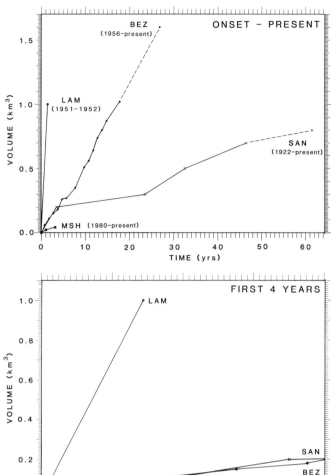

Figure 10. Plots of cumulative seismic energy release, cumulative erupted volume, and cumulative sulfur-dioxide output in metric tonnes at Mount St. Helens during October 1980–December 1983. Plots begin immediately after explosive eruption of October 16–18, 1980, which ended with extrusion of oldest material to survive in current dome. Plots do not include activity or products from explosive eruptions of 1980. Seismic energy curve includes contributions from tectonic-like earthquakes and shallow volcanic earthquakes beneath volcano, plus surface events such as small earthquakes within dome and rock avalanches from its surface. Cumulative sulfur-dioxide output estimated from flux measurements over periods of about 1 hr at intervals of 1 day to about 2 weeks. Dome volumes estimated from sequential topographic maps and subject to revision. Note decline in overall magma supply at the end of 1981 (from about 1.8×10^6 m^3/month to about 0.9×10^6 m^3/month) and approximate volume predictability of activity.

Figure 11. A: Cumulative volume vs. time for lava domes at Mount Lamington (LAM), Bezymianny (BEZ), Santiaguito (SAN), and Mount St. Helens (MSH), starting at onset of dome growth. Mount Lamington data from Taylor (1956, 1958); Bezymianny data from Kirsanov (1979); Santiaguito data from Rose (1973); Mount St. Helens data (simplified) from Table 1. Volumes in 1982 for Bezymianny and Santiaguito extrapolated from earlier data. Seleznev and others (1984) gave a volume of 0.363 km^3 for the Bezymianny dome in 1976, about 1/3 that shown in the figure. Their text is unclear, but their volume appears to be that of the topographic feature reaching above the crater fill, whereas Kirsanov (1979) includes the volume of the dome buried by ejecta and talus. B: Cumulative dome volume vs. time for first four years of dome growth at Mount Lamington, Santiaguito, Bezymianny, and Mount St. Helens. Mount Lamington dome grew to essentially its final volume in about 18 months; others were still growing in August 1984.

grown about 5 times slower than the Bezymianny dome and about 50 times slower than the Mount Lamington dome. Silica and volatile contents, which commonly vary inversely, strongly influence magma viscosity; this comparison suggests that viscosity sometimes is important in governing the maximum rate of dome growth. If other conditions are equal, relatively volatile-rich mafic magma is more mobile than silicic magma and can form domes more rapidly.

THE FUTURE

The capacity of the Mount St. Helens dome to accommodate newly supplied magma without rupturing increases as the dome enlarges, so the ratio of endogenous to exogenous growth will probably continue to increase, unless part or all of the dome

TABLE 2. SELECTED INFORMATION ON FOUR RELATIVELY LONG-LIVED CONTEMPORARY LAVA DOMES

	Mount St. Helens	Mount Lamington[1]	Bezymianny[2]	Santiaguito[3]
Period of Dome Growth				
From	October 1980	February 1951	April 1956	June 1922
To	December 1983	June 1952(?)	December 1982	December 1982
Duration (yr)	3.2+	1.4(?)	27.3	61.5
Dome volume (km^3) (December 1983)	0.044	1.0	1.6	0.8
Average growth rate (km^3/yr)	0.014	0.70	0.06	0.013
Growth rate, first 3 years (km^3/yr)	0.014	0.70 (first 1.4 yrs)	0.06	0.06
SiO_2 content of dome materials	61–64%	58–60%	56–60%	62–65%

[1]Taylor (1958, p. 78) reported volume for Lamington dome at the end of 1955 at 1 km^3, but growth after June 1952 was "negligible" (Taylor, 1956, p. 86). For the purposes of this table and Figure 11, we assume that the dome grew to a volume of 1 km^3 during February 1951–June 1952.

[2]Data from Kirsanov (1979).

[3]Data from Rose (1972, 1973).

is exploded away. Endogenous growth will be especially favored when magma supply is continuous and slow. Extrusive activity will be favored when the dome is unable to accommodate the influx of magma internally, a situation that could result if the short-term supply rate were to exceed the current rate of about 0.9×10^6 m^3/month.

Explosive eruptions are possible, and perhaps even likely, in view of the volcano's history and the histories of similar long-lived domes at other volcanoes. The potential for a large avalanche from the dome to trigger rapid decompression of the magma column and a resulting explosion increases as the dome grows larger. In addition, the dome provides a growing lithostatic load on the magma column, increasing the possibility that magma pressure will eventually reach explosive potential (this effect is relatively small—the dome at the end of 1983 was only 244 m high).

We view the dome, its feeding conduit, and its magma reservoir to be in a delicate balance, able to alternate over short periods of time between dominantly endogenous and exogenous growth and between dominantly explosive and effusive activity.

Careful monitoring will continue to provide the best indicator of future activity at Mount St. Helens, and additional insight can be gained through comparison with similar domes worldwide. The story of the growing lava dome has been colorful so far, but the final chapter has yet to be written.

ACKNOWLEDGMENTS

Many individuals contributed to the observations on which this paper is based. For brevity, we have arbitrarily restricted authorship to those who made significant field observations during two or more episodes of dome growth. Our initial intent was to credit authorship to the "Staff of CVO," but reviewers convinced us that names were desirable. Of the authors, Swanson and Holcomb provided most of the field observations, Swanson wrote initial descriptions of most growth episodes in the form of monthly reports, and Dzurisin wrote the drafts of this paper; all authors revised the drafts and made numerous other vital contributions.

REFERENCES CITED

Bogoiavlenskaya, G. E., and Kirsanov, I. T., 1981, Twenty-five years of activity of Bezymiannyi volcano: Vulkanologiia i seimologiia, no. 1981-2, p. 3–13 (translated by D. B. Vitaliano, November 1981).

Casadevall, T. J., Johnston, D. A., Harris, D. M., Rose, W. I., Jr., Malinconico, L. L., Stoiber, R. E., Bornhorst, T. J., Williams, S. N., Woodruff, Laurel, and Thompson, J. M., 1981, SO_2 emission rates at Mount St. Helens from March 29 through December, 1980, in Lipman, P. W., and Mullineaux, D. R., eds., The 1980 eruptions of Mount St. Helens, Washington: U.S. Geological Survey Professional Paper 1250, p. 193–200.

Casadevall, T. J., Rose, W. I., Gerlach, T. M., Greenland, L. P., Ewert, J., Wunderman, R., and Symonds, R., 1983, Gas emissions and the eruptions of Mount St. Helens through 1982: Science, v. 221, p. 1383–1385.

Cashman, K. V., and Taggart, J. E., 1983, Petrologic monitoring of 1981 and 1982 eruptive products from Mount St. Helens: Science, v. 221, p. 1385–1387.

Chadwick, W. W., Swanson, D. A., Iwatsubo, E. Y., Heliker, C. C., and Leighley, T. A., 1983, Deformation monitoring at Mount St. Helens in 1981 and 1982: Science, v. 221, p. 1378–1380.

Christiansen, R. L., and Peterson, D. W., 1981, Chronology of the 1980 eruptive activity, in Lipman, P. W., and Mullineaux, D. R., eds., The 1980 eruptions of Mount St. Helens, Washington: U.S. Geological Survey Professional Paper 1250, p. 17–30.

Dzurisin, D., Westphal, J. A., and Johnson, D. J., 1983, Eruption prediction aided by electronic tiltmeter data at Mount St. Helens: Science, v. 221, p. 1381–1383.

Gorshkov, G. S., 1959, Gigantic eruption of the volcano Bezymianny: Bulletin Volcanologique, ser. II, v. 20, p. 77–109.

Heliker, C. C., 1984, Inclusions in the 1980–83 dacite of Mount St. Helens, Washington [M.S. thesis]: Bellingham, Western Washington University, 185 p.

Kirsanov, I. T., 1979, Extrusive eruptions on Bezymiannyi volcano in 1965–1974 and their geologic effect: Moscow, Akademiya Nauk SSSR, Sibirskoye Otdeleniye, Institut Geologii i Geofiziki, Nauku, p. 50–69 (translated by D. B. Vitaliano, March 1982).

Malone, S. D., Boyko, C., and Weaver, C. S., 1983, Seismic precursors to the Mount St. Helens eruptions in 1981 and 1982: Science, v. 221, p. 1376–1378.

Moore, J. G., Lipman, P. W., Swanson, D. A., and Alpha, T. R., 1981, Growth of lava domes in the crater, June 1980–January 1981, in Lipman, P. W., and Mullineaux, D. R., eds., The 1980 eruptions of Mount St. Helens, Washington: U.S. Geological Survey Professional Paper 1250, p. 541–548.

Newhall, C. G., and Melson, W. G., 1983, Explosive activity associated with the growth of volcanic domes: Journal of Volcanology and Geothermal Research, v. 17, p. 111–131.

Rose, W. I., Jr., 1972, Santiaguito volcanic dome, Guatemala: Geological Society of America Bulletin, v. 83, p. 1413–1434.

—— , 1973, Pattern and mechanism of volcanic activity at the Santiaguito volcanic dome: Bulletin Volcanologique, v. 37, p. 73–94.

Seleznev, B. V., Dvigalo, V. N., and Gusev, N. A., 1984, Evolution of Bezymiannyi volcano from stereoscopic plotting of aerial photographs of 1950, 1967 and 1976–1981: Volcanology and Seismology, v. 5, p. 53–66.

Swanson, D. A., Casadevall, T. J., Dzurisin, D., Malone, S. D., Newhall, C. G., and Weaver, C. S., 1983, Predicting eruptions at Mount St. Helens, June 1980 through December 1982: Science, v. 221, p. 1369–1376.

Swanson, D. A., Casadevall, T. J., Dzurisin, D., Holcomb, R. T., Newhall, C. G., Malone, S. D., and Weaver, C. S., 1985, Forecasts and predictions of eruptive activity at Mount St. Helens, 1975–1984: Journal of Geodynamics, v. 3, p. 397–423.

Taylor, G.A.M., 1956, An outline of Mount Lamington eruption phenomena: Proceedings, Pan-Pacific Science Congress, 8th, Manila, 1953, v. 2 (Geology and Geophysics; Meteorology), p. 83–88.

—— , 1958, The 1951 eruption of Mount Lamington, Papua: Australian Bureau of Mineral Resources Bulletin, v. 38, 117 p.

Waitt, R. B., Jr., Pierson, T. C., MacLeod, N. S., Janda, R. J., Voight, B., and Holcomb, R. T., 1983, Eruption-triggered avalanche, flood, and lahar at Mount St. Helens—Effects of winter snowpack: Science, v. 221, p. 1394–1397.

Weaver, C. S., Zollweg, J. E., and Malone, S. D., 1983, Deep earthquakes beneath Mount St. Helens: Evidence for magmatic gas transport?: Science, v. 221, p. 1391–1394.

MANUSCRIPT ACCEPTED BY THE SOCIETY MAY 5, 1986

Geological Society of America
Special Paper 212
1987

Volcanic activity at Santiaguito volcano, 1976–1984

William I. Rose
Michigan Technological University
Houghton, Michigan 49931

ABSTRACT

Santiaguito volcano, active for the past 62 yr, has produced almost 1 km^3 of compositionally uniform, soda-rich, calc-alkalic dacite lava that is preserved in 22 distinct extrusive units. The volcanic activity has consisted of endogenous extrusions, block-lava flows, Merapi-type block and ash flows, vertical pyroclastic eruptions, and lahars. Since 1960, block-lava flows have been volumetrically more important than endogenous domes, which were dominant in earlier periods. A 10–12-yr cyclic pattern of activity is marked by the alternation of 3–5 yr periods of high extrusion rates with longer periods of low extrusion rates. The most recent period of high extrusion rates (1972–75), was marked by several large lava flows accompanied by large block and ash flows. During the high extrusion rate period, two distinct vents erupted. However, since 1977 only the Caliente vent, the principal vent at Santiaguito, has remained active. From 1975 through 1984, vertically directed pyroclastic eruptions were especially conspicuous. These erupted bread-crust blocks and locally dispersed fine phreatomagmatic ash. Since 1979, small Merapi-type block and ash flows have been erupted repeatedly from the toe of a small, 400-m-long block-lava flow that is being extruded from the Caliente vent and is flowing down a steep southern slope. On the basis of past behavior, a new period of high extrusion rates is overdue. It may cause an increased extrusion rate for the block-lava flow, or it may result in the extrusion of a new lateral dome unit. At present, the most serious hazards are from mudflows, which occur in each monsoon season and which affect areas more than 10 km from the dome. This problem is worsened by the mid-1970's shift of extrusion to the east side of the dome, which affects river valleys that have been resettled during the past 40 yrs. If the next expected high extrusion rate period focuses on the Caliente vent, as the trends now suggest, the hazard could increase markedly. The last extrusion rate maximum of this type, in 1929–33, resulted in very large block and ash flows that caused many deaths.

INTRODUCTION

Thirty to forty thousand years of basaltic andesite volcanism at Santa María Volcano built the remarkably symmetrical composite cone (volume of 20 km^3) which constitutes the most prominent physiographic feature of the area (Rose and others, 1977b). A repose of at least several hundred years was broken in 1902 with one of the world's largest historic eruptions (Sapper, 1904; Rose, 1972a; Williams and Self, 1983). This Plinian event erupted about 5 km^3 of dacite (dense-rock equivalent) in a 36-hr period, devastated much of southwestern Guatemala, killed several thousand people, contributed to a three-year worldwide average decrease in solar radiation (Humphreys, 1940), and left a gaping amphitheater on the southwest side of Santa María's former symmetrical cone (Fig. 1). In 1922, after 20 yr of relative quiet, lava extrusion began in the center of the 1902 crater; the dome was called Santiaguito (Sapper, 1926). Continued extrusion throughout the succeeding 62 yr has resulted in the formation of a complex dome with 22 distinct units (Rose, 1972b; Rose and others, 1977a). The most dramatic eruption of Santiaguito was a pyroclastic flow erupted in early November 1929 that killed hundreds of people and devastated several villages and plantations (Sapper and Termer, 1930). Santiaguito's magma is compositionally similar to the 1902 dacite, but slightly more

17

Figure 1. Oblique aerial photograph looking north showing Santiaguito, left, and Santa María, right. A typical vertical eruption cloud is shown emerging from the Caliente vent (14 February 1980). The elevation of the summit of Santa María is 3770 m; Santiaguito, 2500 m.

mafic. The emergence of 5 km^3 of 1902 dacite and the nearly 1 km^3 of Santiaguito dacite has converted Santa María to a compositionally bimodal volcano, with a marked compositional gap in the range of 55–62% SiO$_2$. The activity at Santiaguito has been watched sporadically throughout its entire history (Sapper, 1926; Sapper and Termer, 1930; Termer, 1934; Reck and von Tuerkheim, 1936; Termer, 1964; Stoiber and Rose, 1969; Rose and others 1970; Rose, 1972a; Rose and others, 1977b). This paper is the latest in this sequence, and is intended to update the activity at Santiaguito through the end of 1984.

Santiaguito's activity has stimulated a variety of scientific studies in addition to those chronicling its behavior, including work on high-temperature fumarolic incrustations (Stoiber and Rose, 1969, 1974), volcanic-ash particles (Heiken, 1972, 1974; Rose and others, 1980), rate of SO$_2$ emissions from active volcanoes (Stoiber and others, 1983), sedimentary environment of downslope areas of active volcanoes (Kuenzi and others, 1979), and organic compounds in fumarolic gases (Stoiber and others, 1971).

The highland tropical-monsoon climate of Guatemala's coastal slope gives Santiaguito nearly omnipresent fog. In addition, the volcano lies topographically below the volcanic front which precludes its regular observation from an inhabited location and makes it difficult to observe without a specific expedition. Therefore, these summaries are based on only a few days per year of observations. However, the persistence and variety of Santiaguito's activity and the serious volcanic hazards that have been the hallmark of its history are the impetus for assembling the history of its evolution, even though it is fragmentary.

IDEALIZED ERUPTIVE CYCLE OF SANTIAGUITO

Santiaguito has grown in six 3–5-yr spurts of extrusion, which were spaced at 10–12-yr intervals (Fig. 2). Some general characteristics in the eruptive cycle of Santiaguito are: (1) the cycle lasts from 6 to 15 yr, (2) a short (3–5 yr) spurt of high extrusion is interspersed with a longer period of low extrusion (Fig. 2), (3) the high extrusion rate has often been manifested by extrusion at lateral vents west of the Caliente vent, and (4) dome extrusion and lava flows have tended to alternate throughout Santiaguito's history, but both types of extrusion occur in most spurts (Fig. 2). Four of the eruptive events were correlated mainly with the formation of a spatially distinct endogenous dome unit, whereas the other two were correlated with block-lava flows.

During the eruptive spurts of 1939–42, 1949–55, and 1959–63, extrusion focused on three vent areas; La Mitad, El Monje, and El Brujo, which are located along a west-trending line from the Caliente vent. With each new increase in extrusion, the focus shifted westward. I have called the La Mitad, El Monje, and El Brujo areas lateral vents, reflecting their inconsistent, short-lived character. Because of its location within the 1902 explosion crater, I consider the Caliente vent to be the principal or central vent at Santiaguito where magma rises from depth.

The most recent of Santiaguito's eruptive spurts occurred from 1972–1975, and resulted in six block-lava-flow units and a small endogenous domal unit (Figs. 3, 4, and 5). The 1972–1975 spurt was unusual because it added significant extrusive material to Santiaguito from two distinct vents simultaneously. The El Brujo vent, which is located at the west end of the complex, remained active until about 1977. Extrusion from the Caliente vent, located at the eastern end of the complex, began in 1973 and is still continuing. The periods of extrusive activity at Santiaguito's vents is depicted in Figure 2. Much of the activity of the last spurt was summarized by Rose and others (1977a). Several points are worth reemphasizing: (1) The prominence of block-lava flows was significantly more than most previous spurts, and represents a trend started in spurt 5 (1958–63). (2) The addition of large new extrusive units from the Caliente vent was a new development. (3) The occurrence of moderately large pyroclastic flows in 1973 (Rose, 1973b; Rose and others, 1977b) was another new development.

ACTIVITY SINCE 1975

The rate of lava extrusion declined at Santiaguito after 1975, but it did not stop. At the El Brujo vent, lava flow activity stopped and exogenous dome extrusion began. All activity at El Brujo ceased about 1977. The Caliente vent remained active throughout the period of 1973 to 1984. As the overall extrusion rate of Santiaguito declined, explosive vertical eruptions became very prominent and consistent events, especially since 1975. Table 1 is a compilation of the observed activity at Santiaguito since mid-1976. Nearly all of these observations report the occurrence of vertical pyroclastic eruptions from the Caliente vent (Fig. 1). The eruptions, which occur at intervals of 30 minutes to several hours, are generally a few minutes in length. They produce very fine-grained ash, occasionally produce accretionary lapilli, sometimes eject bread-crust bombs of up to 1 m in diameter, and several times each year cause significant ash fall in inhabited areas. These eruptions emit very little SO_2 and are accompanied by a jet engine–like noise.

Similar eruptions occurred in 1968–1969 at Santiaguito, during the low extrusion period which preceded the last spurt (Rose and others, 1970). I interpret these eruptions as phreatomagmatic, but cannot offer conclusive proof of this assertion. The location of the Caliente vent is in the center of the 1902 explosion crater, a location which is consistent with its being the locus of magma rise from depth. This crater location is also likely to funnel meteoric water to the magma column, because the crater receives heavy rainfall, is not drained by rivers, and has a porous, composite cone structure. The fine overall grain size of the ash (Self and Sparks, 1978) and the morphology of the ash particles themselves (Heiken, 1972, 1974) suggests a phreatomagmatic origin. In addition, the low SO_2 content of the eruption clouds (Stoiber and others, 1983) is consistent with this idea.

The volume of ash produced by these eruptions is difficult to determine because the ash is dispersed widely in very thin beds. A cone of debris has been built to a thickness of 50 m around the vent; the debris covers the rough talus surface of the summit of the dome with a smooth carapace. The minimum volume of ash deposited close to the vent since 1975 is about 0.01 km³. Considering that some ash was carried to great distances on occasion (Table 1), the actual volume was probably at least double or triple this minimum. However, when averaged over the nine year period, this volume does not represent a significant contribution to the overall extrusion pattern of Santiaguito.

Beginning in 1975, the cone around the Caliente vent was breached to the south, and a block-lava flow began to emerge (Fig. 6). This lava flow remains as a coherent unit for about 400 m south of the vent, at which point the slope over which it is flowing greatly steepens. Here the flow front continually oversteepens, developing rockfalls of incandescent lava and/or block and ash flows (Fig. 7). This type of activity is termed "Merapi-type" (Williams and McBirney, 1979, p. 152–153). Most of the observations recorded since 1975, particularly those since 1978, have described Merapi-type activity at Santiaguito (Table 1). It probably has continued unabated throughout the entire period.

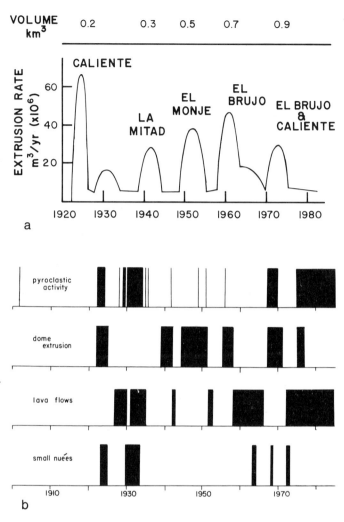

Figure 2. a: Plot showing the estimated variation of magma extrusion rates at Santiaguito as a function of time since 1922. The approximate volume of the dome complex at various times is listed above. The six extrusive spurts discussed in the text are shown. The spurts mainly associated with extrusion at a particular vent are labeled with the name of that vent. The sixth spurt, discussed in the text, was marked by extrusion at both the El Brujo and Caliente vents. b: Schematic representation of the timing of different types of activity at Santiaguito, without any attempt to show the relative scale of different types of activity. Both a and b are updated versions of diagrams from Rose, 1973a. The sources of data are given for the 1922–72 period in Rose, 1973a, and the sources of data for updating are Rose and others, 1977b and this paper.

As of early 1984, this flow represents the only active extrusion at Santiaguito. Thus the active extrusion has shifted from the El Brujo vent back to Caliente. The significance of this shift is not clear. It could represent a transition following the closing of the El Brujo vent, similar to what occurred at the La Mitad vent in the early 1940s and at the El Monje vent in the 1950's (Fig. 2). The renewal of extrusion at the Caliente vent at the end of the 1972–75 spurt, and its continuation since then, could be an important signal that the activity of the volcano will shift completely

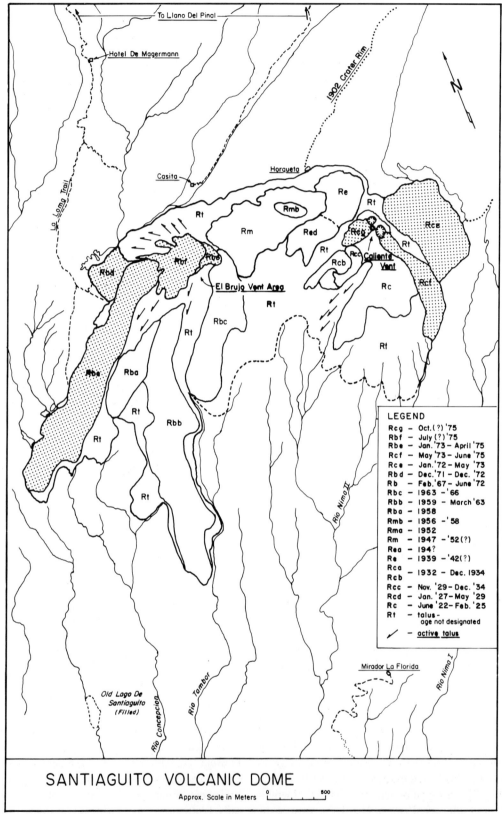

Figure 3. Map of the individual units of the Santiaguito dome, showing individual units and the dates of their extrusion. The units extruded in the 1972–75 spurt are shaded, and the locations of the El Brujo and Caliente vents are shown. Symbols used are consistent with previous papers (Rose, 1972b, 1973a; Rose and others, 1977a).

Figure 4. Oblique aerial photograph of Santiaguito looking north, taken 30 January 1983 by Maurice Krafft. Many of the individual dome units shown in Figure 3 can be recognized. The vertical scale from the summit of Santiaguito to the summit of Santa María is about 1300 m.

Figure 5. Oblique aerial photograph of the El Brujo vent area, looking east-northeast, taken 30 January 1983 by Maurice Krafft. The block-lava flows and late-stage endogenous unit at El Brujo can be seen, and the Caliente vent is visible at the upper right. The horizontal distance from the El Brujo vent to the terminous of the prominent lava flow at left is about 2.5 km (see Fig. 3).

TABLE 1. OBSERVATIONS OF VOLCANIC ACTIVITY AT SANTIAGUITO,
AUGUST 1976-NOVEMBER 1984

Date	SEAN Bulletin*	Comments
Nov-Dec 76	v. 1, no. 14, p. 3	Vertical ash eruptions from Caliente Vent; rockfalls off El Brujo Dome.
25 Jan- 10 Feb 77	v. 2, no. 2, p. 5	Intensification of ash emissions. Significant ashfall at coastal points 70 km SW and at Quezaltenango (12 km NNE). Ash fell for many days and visibility was reduced. Largest eruption was 8 Feb 77.
21 Feb 77	v. 2, no. 5, p. 4	Large vertical eruption and widespread ashfall.
7-19 Mar 77	v. 2, no. 5, p. 4	Large ash eruptions and ashfalls. Top of eruption column reached 6000 m above vent on 19 Mar. El Brujo vent is only weakly active.
23 July 78	v. 3, no. 9, p. 7	Probable block-and-ash flow activity and associated laharic deposition produces disturbance of river channels of the Rio Nima I, Rio Nima II, and Rio Tambor, and subsequent lahars damage farms and bridges.
2 Sept 78	v. 3, no. 9, p. 7	Lahar caused one death and further damage.
Nov-Dec 78	v. 3, no. 11, p. 9	No activity at El Brujo vent. Caliente vent surrounded by 50-m-high rim of ash, blocks and bombs. Vertical eruptions from Caliente vent at 1 to 2-hr intervals, with column tops 1 km above vent. Burned vegetation extends 2 km south of the Caliente vent, indicating recent block-and-ash flows.
23 Aug 79	v. 4, no. 8, p. 15	Seismic disturbance followed by ashfalls at Quezaltenango.
Nov 79	v. 4, no. 11, p. 10-11	Vertical eruptions lasting 3 min at Caliente vent, at 30-min ($\pm$24 sec) intervals to heights of 1.5-1.9 km above vent. Cone around Caliente vent breached to south, with viscous lava flow descending 300 m. Merapi-type block and ash flows decend from the foot of the lava flow.
Jan-Feb 80	v. 5, no. 2, p. 3-4	Vertical Caliente vent eruptions at intervals of 30 min to 6 hrs, with cloud heights to 2.5 km above vent. Larger eruptions 22 Jan, 26 Jan, and 6 Feb. 400-m-long lava flow active south of Caliente vent with frequent block-and-ash flows and rockfalls.
Dec 80	v. 5, no. 12, p. 9-10	Vertical Caliente vent eruptions at intervals of 30 min to 4 hrs, with heights of 0.5 to 1.2 km. Merapi-type block-and-ash flows continue.
12 Feb- 2 Mar 81	v. 6, no. 2, p. 9-10, and v. 6, no. 3, p. 14-15	Vertical Caliente vent eruptions at intervals of 0.5 to 5 hrs, to heights of 0.5 to 2 km. Accretionary lapilli falls. Continuous rockfalls down south slope of Caliente vent as the lava flow and block-and-ash flows continue.
10-11 Feb 82	v. 7, no. 2, p. 7	Vertical Caliente vent eruptions at 1 to 2-hr intervals, to heights of 1 km, and usually lasting 2-5 minutes. Continuous lava extrusion feeding the 300-m-long lava flow from the Caliente vent. Avalanching of flow front several times each hour.
26 Aug 82	v. 7, no. 9, p. 12	Lahars along the Rio Nima cause evacuation of hundreds of residents south of Santiaguito.
29-30 Jan 83	v. 8, no. 1, p. 12	Vertical explosions from Caliente vent at 2-hr intervals, heights of 2-3 km above vent. No avalanching of block-and-ash flows noted.
Jun-Aug 83	v. 8, no. 11, p. 9-10	Severe disturbances in the flow of rivers in the vicinity of El Palmar, a village 10 km south of Santiaguito, was caused by laharic deposits related to continued Caliente vent activity. Evacuations of several hundred people and destruction of several dozen homes.
Nov 83	v. 8, no. 11, p. 9-10	Vertical Caliente vent eruptions at 0.5-hr intervals and heights of 0.2-1 km above vent. Incandescent avalanches from the base of the block lava flow south of the Caliente vent.
Nov 84-Feb 85	v. 10, no. 2, p. 9	Large ash eruptions with ash fallout at El Palmar (25 Nov). Vertical eruptions at 0.1-1 hr intervals at 1.2 km height (24-25 Jan). Shallow B-type earthquakes accompany eruptions and generally high seismicity recorded (late Jan-early Feb).

*SEAN Bulletin = Smithsonian Event Alert Network Bulletin, Smithsonian Institution, Washington, D.C., USA.

Figure 6. Oblique aerial view of the Caliente vent looking northeast, taken 7 February 1980. The view shows the breached cone around the vent and a 400-m-long block-lava flow extruding from the vent. This extrusion has continued for nine years, but the flow has not lengthened. Instead, the flow forms avalanches at its foot, resulting in block and ash flows and hot avalanches. (For scale, cf. Fig. 4.)

Figure 7. View of Santiaguito during active avalanching of the flow front shown in Figure 6. a: Oblique aerial view looking north, taken on 14 February 1980. b: View from the summit of the El Monje dome, looking east-southeast to the Caliente vent; the flow slopes to the right (For scale, cf. Fig. 4.)

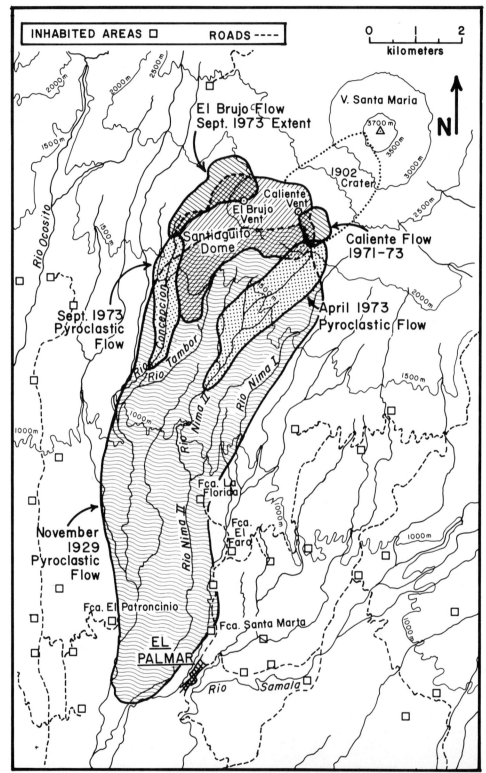

Figure 8. Map showing the areas devastated by large pyroclastic flows at Santiaguito in 1929 and in 1973. The area affected in the 1929 event was constructed from the observations of Ospina (1929). Descriptions of the 1929 event are in Sapper and Termer, 1930. The 1973 events were described by Rose (1973b) and Rose and others (1977a). The extent of active lava flows in 1973 is shown by shading. The September 1973 pyroclastic flow erupted from the foot of the El Brujo flow (Rbe in Fig. 3), and it is possible that the April 1973 pyroclastic flow was related to the shaded Caliente flow (Rce in Fig. 3).

Figure 9. Photo showing a part of El Palmar damaged by mudflows in September 1983.

Figure 10. Photograph of a 1.5-m-thick laharic unit typical of the deposits in the Río Nima II formed during the rainy season of 1983.

back to the east side of the complex. If true, this shift is important to volcanic hazard assessments.

LAHARIC DEPOSITION SOUTH OF SANTIAGUITO

The shift of extrusion from the El Brujo vent back to Caliente has markedly affected the downslope areas. From about 1940 until the early 1970s, relative calm has prevailed on the eastern slopes of the dome complex and the downslope areas have been resettled and partly developed. Thus the renewal of laharic activity in the headwaters of the Río Nimá I and Río Nimá II, as well as the Río Tambor, has caused some loss of life and property. The occurrence of lahars correlates closely with the monsoon rains, which occur from May to November (Table 1).

An example of the type of event which is presently occurring is the lahar sequence of July and August of 1983. A blockage of the headwaters of the Río Nimá II was the trigger for the damaging lahars which affected the town of El Palmar (680 m elevation, 10 km south and 1900 m below Santiaguito (Fig. 8). The town is located at the confluence of the Río Nimá I and the Río Nimá II. Lahars in the Río Nimá II gradually built a dam that created a lake. The lake eventually overflowed its banks triggering lahars that inundated many dozens of houses with laharic material (Fig. 9). In this instance, sufficient time was available so that people could be evacuated in time to avoid injury or death. However, the possibility of lahars in these valleys is a constant threat to people and livestock.

Figure 10 is a photograph of the reverse-graded laharic deposits which characterize these river channels. The rivers alternately cut into their channels and then aggrade them, making accurate forecasting of lahars difficult without regular observations of the channels and the volcano.

THE NEXT SPURT OF EXTRUSION AT SANTIAGUITO

Based on past history (Fig. 2), the seventh spurt of extrusion at Santiaguito can be expected soon. If extrusion rate at the Caliente vent increases (as appears likely), there is a possibility of large block and ash flows. The historic example of most interest is the 2–4 November 1929 event (Sapper and Termer, 1930). The area affected by this eruption is shown in Figure 8. If an event of similar magnitude occurred today, it is likely that the loss of life

Figure 11. Aerial photograph of Santiaguito from the east, taken in March 1933 by O. G. Von Tuerkheim. Steam and gas is being emitted from the Caliente vent and the Merapi-type extrusion from the vent can be seen. Compare with Figure 6.

would exceed that experienced in 1929 because of resettlement and a greater population density. The present activity at Santiaguito resembles the period of 1928–33. An extrusive spurt has ended, vertical explosions are common at the Caliente vent, a block-lava flow is being extruded, and attendant Merapi-type activity is occurring, just as in 1929 (Fig. 11). If a new extrusive spurt were to begin, larger, Pelean-type pyroclastic flows could result.

ACKNOWLEDGMENTS

Financial support for expeditions to Santiaguito came from National Science Foundation Grants DES 78-01190 and EAR 82-05606 and National Aeronautics and Space Administration Grant NAG 1-200. Pedro Perez Lopez accompanied me on all trips to the volcano. Maurice and Katia Krafft supplied me with excellent aerial photographs. Eddy Sanchez, Edgar Quevec, and Carlos Martinez (Instituto Nacional de Sismologia, Vulcanologia, Meteorologia e Hidrologia [Insivumeh], Guatemala) have begun systematic observations and measurements at Santiaguito since 1984 and were helpful in preparing this report. Tom Moyer, Dean Eppler, and Joy Crisp were enthusiastic reviewers.

REFERENCES CITED

Heiken, G., 1972, Morphology and petrography of volcanic ashes: Geological Society of America Bulletin, v. 83, p 1961–1988.
——, 1974, An atlas of volcanic ash: Smithsonian Contributions to Earth Sciences, v. 12, 101 p.
Humphreys, W. J., 1940, Physics of the air: New York, McGraw-Hill, 676 p.
Kuenzi, W. D., Horst, O. D., and McGehee, R. V., 1979, Effect of volcanic activity on fluvial-deltaic sedimentation in a modern arc-trench gap, southwestern Guatemala: Geological Society of America Bulletin, v. 90, p. 827–838.

Ospina, C. W., 1930, Cuatro horas de marcha por el desierto de arena a corta distancia del crater del Santa María: Guatemala, Anales de la Sociedad de Geografia e Historia, v. 7, no. 1, p. 68–79.
Reck, H., and von Tuerkheim, O. G., 1936, Der Zustand der Vulkane Fuego, Atitlán, und Santa María in Guatemala Ende 1934: Zeitschrift für Vulkandogie, v. 14, p. 259–263.
Rose, W. I., 1972a, Notes on the 1902 eruption of Santa María Volcano, Guatemala: Bulletin Volcanologique, v. 36, p. 29–45.
——, 1972b, Santiaguito volcanic dome, Guatemala: Geological Society of

America Bulletin, v. 83, p. 1413–1434.

——, 1973a, Pattern and mechanism of volcanic activity at the Santiaguito volcanic dome, Guatemala: Bulletin Volcanologique, v. 37, p. 73–94.

——, 1973b, Nuée ardente from Santiaguito Volcano, April 1973: Bulletin Volcanologique, v. 38, p. 365–371.

Rose, W. I., Stoiber, R. E., and Bonis, S. B., 1970, Volcanic activity at Santiaguito Volcano, Guatemala, June 1968–August 1969: Bulletin Volcanologique, v. 34, p. 295–307.

Rose, W. I., Pearson, T., and Bonis, S., 1977a, Nuée ardente eruption from the foot of a dacite lava flow, Santiaguito Volcano, Guatemala: Bulletin Volcanologique, v. 40, p. 53–70.

Rose, W. I., Grant, N. K., Hahn, G. A., Lange, I. M., Powell, J. L., Easter, J., and Degraff, J. M., 1977b, The evolution of Santa María volcano, Guatemala: Journal of Geology, v. 85, p. 63–87.

Rose, W. I., Chuan, R. L., Cadle, R. D., and Woods, D. C., 1980, Small particles in volcanic eruption clouds: American Journal of Science, v. 280, p. 671–696.

Sapper, Karl, 1904, Die vulkanischen Ereignisse in Mittelamerika in Jahre 1902: Neues Jahrbuch für Mineralogie Geologie und Paläontologie, v. 1, p. 39–90.

——, 1926, Die Vulkanische Tätigkeit in Mittelamerika im 20 Jahrhundert: Zeitschrift für Vulkanologie, v. 9, p. 156–203.

Sapper, K., and Termer, F., 1930, Der Aüsbruch des Vulkans Santa María in Guatemala vom 2–4 November 1929: Zeitschrift für Vulkanologie, v. 13, p. 73–101.

Self, S., and Sparks, R.S.J., 1978, Characteristics of widespread pyroclastic deposits formed by the interaction of silicic magma and water: Bulletin Volcanologique, v. 41, p. 196–212.

Stoiber, R. E., and Rose, W. I., 1969, Recent volcanic and fumarolic activity at Santiaguito Volcano, Guatemala: Bulletin Volcanologique, v. 33, p. 475–502.

——, 1974, Fumarole incrustations at active Central American volcanoes: Geochemica et Cosmochimica Acta, v. 38, p. 495–516.

Stoiber, R. E., Rose, W. I., Leggett, D. C., Jenkins, T. F., and Murrmann, R. P., 1971, Organic compounds in volcanic gas from Santiaguito volcano, Guatemala: Geological society of America Bulletin, v. 82, p. 2299–2302.

Stoiber, R. E., Malinconico, L. L., and Williams, S. N., 1983, Use of the correlation spectrometer at volcanoes, in Tazieff, H., and Sabroux, J. C., eds., Forecasting volcanic events: Amsterdam, Elsevier, p. 425–444.

Termer, Franz, 1934, Die tätigkeit des Vulkans Santa María in Guatemala in den Jahren 1931–33: Zeitschrift für Vulkanologie, v. 16, p. 43–50.

——, 1964, Die tätigkeit der Vulkane von Guatemala in den Jahren 1960–1963: Petermanns Geographische Mitteilungen, v. 108, p. 261–268.

Williams, H., and McBirney, A. R., 1979, Volcanology: San Francisco, Freeman, Cooper and Co., 391 p.

Williams, S. N., and Self, S., 1983, The October 1902 plinian eruption of Santa María Volcano, Guatemala: Journal of Volcanology and Geothermal Research, v. 16, p. 33–56.

MANUSCRIPT ACCEPTED BY THE SOCIETY MAY 5, 1986

Printed in U.S.A.

Eruptive histories of Lipari and Vulcano, Italy, during the past 22,000 years

Michael F. Sheridan
Department of Geology
Arizona State University
Tempe, Arizona 85287

G. Frazzetta
Istituto Internazionale di Vulcanologia, C.N.R.
v.le Regina Margherita 6
95123 Catania
Italy

L. La Volpe
Dipartimento Geomineralogico, Universitá di Bari,
Piazza Umberto I, no. 1,
70121 Bari
Italy

ABSTRACT

Silicic volcanism in the central Aeolian Islands of Lipari and Vulcano has followed a consistent pattern during the past 22,000 years. Active eruptive cycles generally began with hydrovolcanic breccias, surge beds, and ash-fall deposits. They ended with magmatic effusions that formed lava domes and short coulees. Long repose periods separated shorter active cycles. Eruptions occurred from both isolated vents located along fissures (e.g., domes of southern Lipari) and central vents with a long history of activity and repose (e.g., Fossa cone of Vulcano). The compositions of the juvenile products include leucite tephrite, trachyte, and rhyolite.

The average volume of silicic products in an eruptive cycle on Lipari and Lentia was about 5×10^8 m^3 of juvenile magma. The repose period between major active periods was about 4,000 years. The production rate for the period of 22,000 years ago to the present was 10^5 m^3 per year. The average volume of erupted material in an active cycle at Fossa (other than the Punte Nere cycle) was about 2×10^7 m^3 of juvenile magma. Repose times between cycles range from 300 to 800 years. The rate of magma production for the entire Fossa cone during its 6,000-year growth was 5×10^4 m^3 per year. Vulcanello produced about 3×10^7 m^3 of tephritic to trachytic magma in the past 2,100 years, a production rate of 1.5×10^4 m^3 per year.

INTRODUCTION

Lipari and Vulcano are adjacent islands at the center of the Aeolian Archipelago, which is located north of Sicily on the southern margin of the Tyrrhenian Sea (Fig. 1). These islands comprise the summits of a large volcanic edifice, the base of which is on the Tyrrhenian Sea floor at a depth of approximately 2,000 m. Although the magma sources for Lipari, Vulcano, and Vulcanello are independent, a common tectonic regime controls the complex relationship of the feeding systems on the main volcanic edifice. The simultaneous or sequential eruption of vents separated by a few kilometers and the mixture of magmas of various types (tephrite, trachyte, and rhyolite, according to Keller, 1980) support this contention.

The purpose of this paper is: (1) to describe the eruptive and repose histories of Lipari and Vulcano, and (2) to use the volumes and ages of the deposits to constrain the magma production rates. Data for the present analysis is fairly complete for the period dating back to 22,000 B.P. Some of the eruptive events (e.g., the formation of the Lentia dome complex, the Caldera of the Fossa, and the Vulcanello shield) require more detailed study.

Three types of eruptive systems emitted the most recent

volcanic products on Lipari and Vulcano: (1) fissure-related tuff and dome complexes on Lentia and Lipari; (2) complex tuff cone on Fossa of Vulcano; and (3) lava shield of Vulcanello. Each eruptive system type has a coherent pattern of repose and eruption as well as a consistent petrologic character.

METHOD OF STUDY

The objective of our studies on Lipari and Vulcano, which began in 1980, is to understand the nature of volcanic activity for the purpose of evaluating volcanic risk. The basis of our interpretations is a detailed volcanic stratigraphy of the youngest deposits obtained through a bed-by-bed measurement of pyroclastic sections on a centimeter scale. We grouped the eruptive products into volcanic cycles that accumulated during more or less continuous periods of activity. Somewhat longer periods of repose separate the active phases. Interpretation of mechanisms of eruption and emplacement are based on bedding structures, grain-size analysis, and clast morphology.

The numerous stratigraphic sections measured for each deposit allowed the construction of isopach maps for the main pyroclastic units (Monte Guardia sequence, Gabellotto-Fiume Bianco tephra and Pilato tephra on Lipari, and the products of the various eruptive cycles of the Fossa composite cone on Vulcano). The volumes for the various tephra units were calculated by extrapolating the thickness vs. area data to a thickness of 1 cm, by using the method of Rose and others (1983). Volume estimates of tephra could be as much as 50% too low, because extrapolation to thicknesses less than 1 cm was prevented by lack of isopach data. Most of the tephra deposits are of low-energy, hydromagmatic types so that their thickness decreases sharply with distance. This minimizes the error in volume estimates due to uncounted distal ash. The volume of tephra associated with the Vulcanello and Lentia vents is unknown because these volcanoes have not yet been studied in detail.

Volumetric data for the domes and lavas come from measurements of distributions and thickness taken from existing geologic and topographic maps. The areas of the various units were measured using a digitizing tablet and the average thickness estimated from the exposures and the maps. Generally, several domes comprise each cycle on Lentia and Lipari, whereas a single lava flow is characteristic of each cycle of the Fossa cone.

The chronology of events was determined by several techniques. Legends and eye-witness accounts (Frazzetta and others, 1984) were used to date the eruptions that occurred during historical times (approximately the past 2,000 years). Crisci and others (1983) dated carbonized wood from several of the brown ash-flow tuffs of Lipari for the time period of 16,800 to older than 35,000 B.P.

Because there is no carbonized material in the deposits on Lentia or the Fossa of Vulcano, K-Ar dates were obtained from the lavas. The method for dating such young rocks is described elsewhere (Cassignol and others, 1978; Cassignol and Gillot, 1982; Gillot and others, 1982). Although the K-Ar dates have a large standard deviation, their values are consistent with stratigraphic and historic data (Frazzetta and others, 1984). The few widespread ash layers that allow for correlations of deposits on Lipari and Vulcano confirm ages obtained by the various methods of dating.

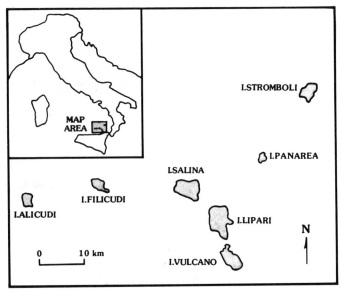

Figure 1. Index map of the Tyrrhenian Sea showing the location of Lipari and Vulcano in the center of the Aeolian Archipelago.

THREE ERUPTIVE SYSTEM TYPES

Lipari-Lentia System

The Lipari-Lentia type system has an eruptive record that dates from before 35,000 B.P. (Pichler, 1980; Crisci and others, 1981). Volcanoes related to this type of system are fed through NNW- and NE-trending fissures (Frazzetta and others, 1984). Each studied eruptive cycle exhibited an initial stage of explosive hydrovolcanism that was followed by the formation of a series of domes or thick lava flows. The common magma composition associated with these vents is rhyolite (Pichler, 1980), although there are some trachytic lavas on Lentia (Keller, 1980). Minor clasts of shoshonitic composition were mixed with the dominantly calc-alkaline rhyolitic products of the Monte Guardia sequence (De Rosa and Sheridan, 1983).

The main eruptive cycles related to this group of vents from oldest to youngest are: (1) a complex of older domes on the south end of Lipari; (2) the Monte Guardia series (about 21,000 B.P.), (3) the Lentia complex on Vulcano (about 15,500 B.P.), (4) the Gabellotto-Fiume Bianco deposits (11,400 to 8,300 B.P.), and (5) the Pilato cycle materials (1,400 B.P.). Eruptions of this type generally open with hydromagmatic or pyroclastic explosions and close with the slow effusion of rhyolitic lava.

The age, stratigraphy, and volume of the older series of domes on Lipari have not yet been studied in detail. Pichler

(1980) interpreted these domes as the first predominantly rhyolitic products of his third period that followed the dominantly andesitic activity of his second period. Preliminary data of Barker (1984) on the Monte San Angelo products contradicts Pichler's interpretation and suggests that some of the older supposed andesitic lavas are actually mixtures of rhyolitic and mafic magmas.

The products of the older group of rhyolites on Lipari form a north-trending chain of partially submerged domes that include Falcone, Capparo, Capistello, and Lipari (Fig. 2.1). The activity of this age was probably similar to that of the subsequent Monte Guardia cycle, but the volume of lava appears to be somewhat greater. The amount and distribution of tephra associated with this activity is not known. The age of these domes, based on tephra stratigraphy of the brown tuffs (Crisci and others, 1983), is more than 23,500 ±900 B.P., but the maximum age is not well constrained (probably about 35,000 B.P.).

Materials of the Monte Guardia sequence (Crisci and others, 1981) are the oldest well-documented products in this group (Fig. 2.2). The basal part of the section is a thick, lithic-rich explosion breccia deposit that is overlain by interbedded pumice-fall layers and surge beds. These tephra contain juvenile clasts of various compositions that suggest mixing of mafic and silicic magma during the eruption (De Rosa and Sheridan, 1983). This pyroclastic phase was followed by the extrusion of the thick domes of San Lazzaro, Monte Guardia, and Monte Giardina (Fig. 2.2) that lie on a north-northwest line. Brown ash-flow tuff horizons above (20,300 ± 700 B.P.) and below (22,480 ±1,100 B.P.) the Monte Guardia sequence products provide tight age constraints on this cycle (Crisci and others, 1983).

The stratigraphy of the Lentia complex on Vulcano has not yet been studied, and no pyroclastic deposits have been identified. The domes are aligned in a NNW direction (Fig. 2.3), similar to the domes of the Monte Guardia sequence. The lavas range in composition from rhyolite to trachyte (Keller, 1980), which is a greater chemical variation than exists among domes of other ages. Radiometric data (K-Ar dates of Frazzetta and others, 1984) give ages of 15,000 ±1,500 and 16,100 ±1,300 B.P. for lavas from Lentia.

Stratigraphic reconstruction of deposits on northern Lipari by Cortese and others (1986) indicates that several eruptions took place from separate vents following the emplacement of the upper ash-flow units (16,800 ±200 B.P.) of Crisci and others (1983). The first phase of this activity produced the Canneto Dentro unit (Cortese and others, 1986), that consists of an explosion breccia and surge beds that are overlain by a small lava flow (Fig. 2.4).

The second phase of this eruptive cycle in northern Lipari produced the Gabellotto-Fiume Bianco unit (Fig. 2.4). The pyroclastic part of this sequence consists of a tuff ring composed of surge beds with large-scale cross-stratification and small pyroclastic-flow deposits (Cortese and others, 1986). The cycle ended with the extrusion of the large Pomiciazzo lava dome which yielded fission-track ages of 11,400 ±1,800 and 8,600 ±1,500 B.P. (Bigazzi and Bonadonna, 1973; Wagner and others, 1976).

A long period of quiescence, marked by a 1.5-m-thick paleosol, occurred before the renewal of activity. Pichler (1980) interpreted this paleosol to represent a time interval of at least 3,500 years, because various ^{14}C ages were determined on carbonized wood from this horizon (from 4,810 ±60 to 1,220 ±100 B.P.).

The last period of activity on Lipari also consists of two phases. The first phase formed the Forgia Vecchia unit (Cortese and others, 1986); a breccia ring that is covered with a stratified, fine-ash deposit. This tephra deposit is capped by the Forgia Vecchia lava flow (Fig. 2.6). The second phase of activity produced the Monte Pilato-Rocche Rosse unit (Cortese and others, 1986). The Monte Pilato deposits are complex, consisting of a large pumice cone and a surrounding blanket of fine ash. The final activity was the extrusion of the Rocche Rosse lava flow. The growth of the pumice cone and eruption of its associated ash is dated as sixth century A.D., on the basis of archeological data (Keller, 1980) and the legend of San Calogero (about A.D. 580). The final phase of lava eruption was observed by San Willibald in A.D. 729 (Bernabo-Brea and Kronig, 1978).

Vulcano

The Fossa cone of Vulcano has a record of activity that dates from 5,500 B.P. (Frazzetta and others, 1984). The complex tuff cone surrounding the central conduit contains only a minor volume of lava (Frazzetta and others, 1983). Juvenile materials consist of early trachytes and later rhyolites (Keller, 1980). Recent products commonly contain mixtures of trachyte melt blobs and stringers in a calc-alkaline rhyolite lava matrix.

The Fossa complex cone developed on the floor of a caldera (Fig. 2.4) that collapsed after the extrusion of the Lentia lavas (15,500 ±1,400 B.P.). The Punte Roja tephritic lava (14,000 ±6,000 B.P.) flowed onto the southeastern floor of the caldera. The Fossa complex cone was formed by at least seven cycles of activity (Figs. 2.5 and 2.6). K-r dates on lavas that end cycles (Frazzetta and others, 1984) include Punte Nere (5,500 ±1,300 B.P.), Campo Sportivo (4,600 ±1,700 B.P.), Palizzi (1,600 ±1,000 B.P.), Commenda (about 1,000 B.P.), and Pietre Cotte (240 B.P.). Deposits of two unspecified cycles have ages that lie sometime between 4,500 and 1,600 B.P. The cycles on Fossa have a pattern of hydromagmatic eruptions followed by magmatic activity. Products from the initial phases of the cycle consist of explosion breccias or wet-surge deposits. The characteristics of the deposit change upward to dry-surge beds, and each cycle ends with a lava flow (Frazzetta and others, 1983).

Vulcanello

Vulcanello is the youngest eruptive system; it has an inception of subaerial activity that dates from historic times (Fig. 2.6). This volcano consists of a low shield of leucite tephrite lava. A scoria-cone nucleus and adjacent tuff cones form the central vent. Some shoshonitic pyroclastic materials on Lipari may be related

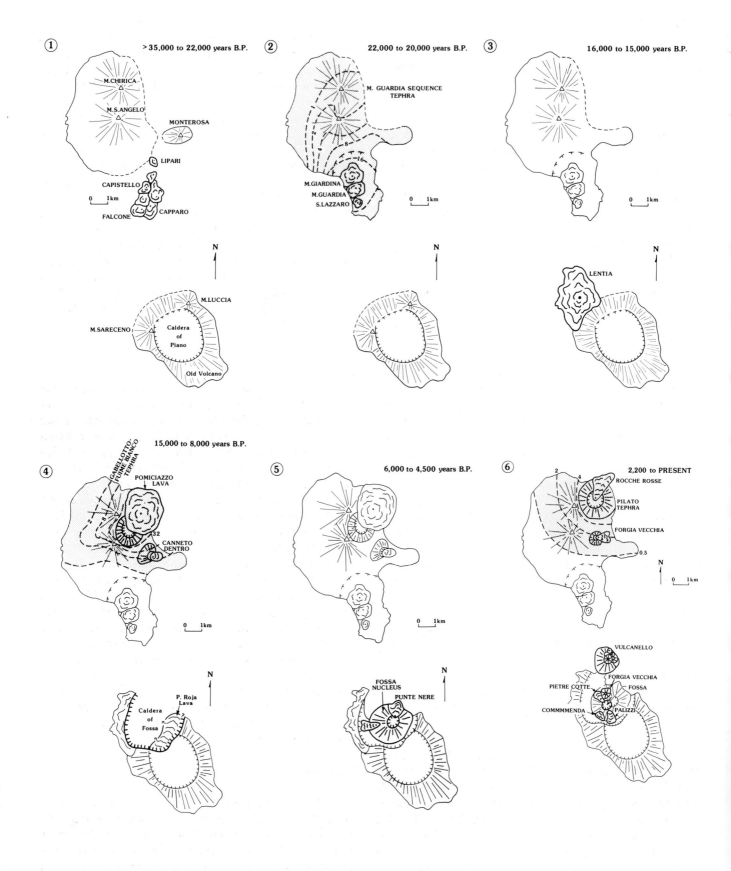

to the Vulcanello system. Brown ash-flow tuffs on Lipari that date from 16,800 to older than 35,000 B.P. may have erupted from the Vulcanello vent system before the edifice reached the surface of the ocean (Crisci and others, 1983).

The stratigraphy of Vulcanello has not yet been studied in detail; therefore, an accurate designation of eruptive cycles and repose periods can not be made. Plinius and Polibius (cited in Mercalli, 1891) recorded the birth of Vulcanello as 183 B.C. In contrast, Strabo (cited in Mercalli and Silvestri, 1891) suggested 126 B.C. for the appearance of the island and reported activity in 91 B.C. and formation of the lava platform of Vulcanello by the end of the first century B.C. According to the legend of San Cologero, Vulcanello and Fossa exhibited contemporaneous activity during the sixth century; this activity suggests a common triggering mechanism for these events. Vulcanello was probably active at intermediate times, but the products of a large eruption in 1550 created the isthmus connecting it to Vulcano (De Fiore, 1922; Mercalli and Silvestri, 1891).

DISCUSSION

Existing data allow some conclusions to be drawn about the eruption rates for the various venting systems on Lipari and Vulcano. All of the eruptive cycles of the Lipari-Lentia type systems that were examined in detail began with a hydrovolcanic or pyroclastic phase and ended with the extrusion of lava flows or domes. Several domes were produced in the earlier cycles but the later cycles erupted only a single dome or lava. Except for the Monte Guardia series, the volume of final lava far exceeds that of the pyroclastic materials (Fig. 3). Therefore, cumulative volume calculations are probably an accurate estimate of erupted magma.

The active periods were much shorter than the repose times. The average repose time between the main active phases is about 4,000 yr. The volume of erupted material in each cycle is between 2×10^8 and 8×10^8 m^3. The time vs volume plot for these vents (Fig. 4) indicates an average rate of lava production of 10^5 m^3 per year.

Age and volume data are available for several cycles of activity on the Fossa complex cone of Vulcano (Fig. 5). Lavas comprise only a small percentage of the products of each cycle (between 1% and 10%). Therefore, the volume calculations depend on accurate estimates of tephra distributions. Because the explosive products were largely hydromatic (Sheridan and Wohletz, 1983), the pyroclastic units tend to form cones. Their volume of distal ashes is probably small, but possibly significant.

The Punte Nere cycle, which formed the nucleus of the Fossa cone complex, was by far the most voluminous of the eruptions (Fig. 5). The volume of products for the trachytic cycles (Punte Nere through Palizzi) decreases with time. The Com-

Figure 2 (facing page). Geomorphic development of Lipari and Vulcano during the past 35,000 years. 2.1: The older rhyolitic domes were extruded at the south end of Lipari between 35,000 and 22,000 B.P. 2.2: About 21,000 B.P., the Monte Guardia tephra on Lipari were catastrophically emplaced. Extrusion of the domes of Monte Giardina, Monte Guardia, and San Lazzaro followed. Isopachs of tephra thickness (in meters) shown by dashed lines. The location of a possible caldera is marked by the concentric symbols placed along an escarpment to the northwest of Monte Giardina. 2.3: The Lentia complex of rhyolitic and trachytic domes was emplaced to the northwest of the old Vulcano structure at approximately 15,500 ±1,400 B.P. 2.4: Between 15,500 and 14,000 ±6,000 B.P., the caldera of Fossa on Vulcano collapsed and the floor was partially covered with the tephritic Punte Roja lava. In the time interval of 11,000 to 8,300 B.P., the Canneto Dentro and Gabellotto-Fiume Bianco units were emplaced on Lipari. Each produced a tephra deposit; extrusion of lava followed. Tephra units are shown by stippled pattern and the thickness is given in meters. 2.5: The Punte Nere cycle ended at 5,500 B.P.; it formed the nucleus of the Fossa cone complex. The Campo Sportivo lava (4,500 B.P.) and tephra from at least two other undefined cycles make up the early deposits of this central volcano. 2.6: The past 2,200 years were marked by the eruption of the Forgia Vecchia and Rocche Rosse lavas and the associated Pilato tephra on Lipari. The tephra deposit is shown by a stippled pattern and the thickness is given in meters. The Fossa complex cone of Vulcano records the Palizzi, Commenda, and Pietre Cotte cycles, each of which ended with a lava flow. During this period, the small central volcano of Vulcanello was born.

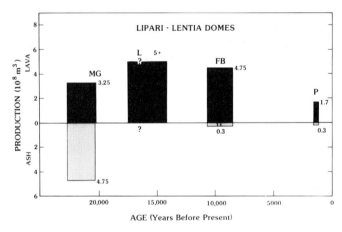

Figure 3. The relationship of volume of erupted products and age for the domes of Lentia and Lipari. MG = Monte Guardia sequence; L = Lentia complex; FB = Gabellotto-Fiume Bianco unit; P = Pilato unit.

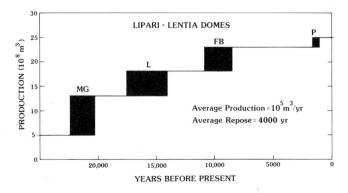

Figure 4. Production rate for the fissure-related domes of Lentia and Lipari. Units as in Figure 3.

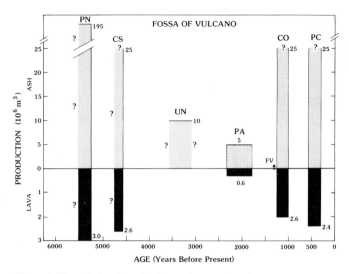

Figure 5. The relationship of volume of erupted products and age for the Fossa of Vulcano. PN = Punte Nere cycle; CS = Campo Sportivo cycle; UN = unnamed cycles; PA = Palizzi cycle; CO = Commendo cycle, PC = Pietre Cotte cycle.

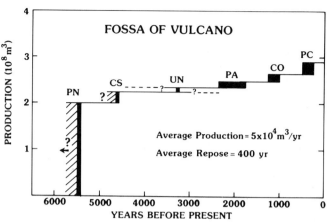

Figure 6. Cumulative production and average production rate on Fossa of Vulcano. Units as defined in Figure 5.

menda and Pietre Cotte rhyolitic cycles produced a slightly larger volume of material than the late trachytic eruptions. The eruptive periods on Fossa lasted for about 200 yr; repose periods lasted 300 to 800 yr. The magma eruption rate for the Fossa vent is 5×10^4 m^3 per year (Fig. 6).

There are insufficient stratigraphic and areal data to determine the distribution of erupted products of Vulcanello in time and space. The volume of the present cone above sea level yields a production rate of 1.5×10^4 m^3 per year for the past 2,100 years.

ACKNOWLEDGMENTS

This work was supported by National Science Foundation Grant EAR 8306168 and by the Consiglio Nazionale delle Ricerche of Italy. An earlier version of the manuscript was improved by the careful review of James Luhr, Thomas Moyer, Eugene Smith, and Thomas Vogel.

REFERENCES CITED

Barker, D. S., 1984, High-potassium S-type rhyolite, Monte S. Angelo, Lipari, Aeolian Islands, Italy: Geological Society of America Abstracts with Programs, v. 16, no. 6, p. 437.

Bernabò-Brea, L., and Krönig, W., 1978, Le Isole Eolie dal tardo antico ai Normanni: Archivio Storico Siracuso, nuova ser. 5, p. 1–99.

Bigazzi, G. and Bonadonna, F., 1973, Fission track dating of the obsidian of Lipari Island (Italy): Nature, v. 242, p. 322–323.

Cassignol, C., and Gillot, P.-Y., 1982, Range and effectiveness of unspiked potassium-argon dating, in Odin, G. S., ed., Numerical dating in stratigraphy: New York, Wiley and Sons, p. 160–179.

Cassignol, C., David, B., and Gillot, P.-Y., 1978, Technologie potassium-argon: Rapport CEA-R-4908 CEN Asclay.

Cortese, M., Frazzetta, G., and LaVolpe, L., 1986, Volcanic history of Lipari (Aeolian Islands, Italy) during the last 10,000 years: Journal of Volcanology and Geothermal Research, v. 27, p. 117–133.

Crisci, G. M., DeRosa, R., Lanzafame, G., Mazzuoli, R., Sheridan, M. F., and Zuffa, G. G., 1981, Monte Guardia sequence: A late Pleistocene eruptive cycle on Lipari (Italy): Bulletin Volcanologique, v. 44, p. 241–255.

Crisci, G. M., Delibrias, G., DeRosa, R., Mazzuoli, R., and Sheridan, M. F., 1983, Age and petrology of the late Pleistocene brown tuffs on Lipari, Italy: Bulletin Volcanologique, v. 46, p. 381–391.

De Fiore, 1922, Vulcano (Isole Eolie): Zeitschrift für Vulkanologie, v. 3, p. 3–393.

De Rosa, R., and Sheridan, M. F., 1983, Evidence for magma mixing in the surge deposits of the Monte Guradia sequence, Lipari: Journal of Volcanology and Geothermal Research, v. 17, p. 313–328.

Frazzetta, G., La Volpe, L., and Sheridan, M. F., 1983, Evolution of the Fossa cone, Vulcano: Journal of Volcanology and Geothermal Research, v. 17, p. 329–360.

Frazzetta, G., Gillot, J. P., La Volpe, L., and Sheridan, M. F., 1984, Volcanic hazards at Fossa of Vulcano: data from the last 6,000 years. Bulletin Volcanologique, v. 47, p. 105–124.

Gillot, P.-Y., Chiesa, S., Pasquare, G., and Vezzoli, L., 1982, >33,000-yr K-Ar dating of the volcano-tectonic horst of the Isle of Ischia, Gulf of Naples: Nature, v. 299, p. 242–244.

Keller, J., 1980, The island of Vulcano: Rendiconti della Società Italiana di Mineralogia e Petrologia, v. 36, p. 369–414.

Pichler, H., 1980, The island of Lipari: Rendiconto della Società Italiana di Mineralogia e Petrologia, v. 36, p. 415–440.

Mercalli, G., and Silvestri, O., 1891, Le eruzioni dell'isola di Vulcano, incominciate il 3 agosto 1888 e terminante il 22 marzo 1890. Relazione scientifica, 1891: Annuo Ufficio Centrale Meteorologia y Geodinamica, v. 10, no. 4, p. 1–213.

Rose, W. I., Wunderman, R. L., Hoffman, M. F., and Gale, L., 1983, A volcanologist's review of atmospheric hazards of volcanic activity: Fuego and Mount St. Helens: Journal of Volcanology and Geothermal Research, v. 17, p. 133–157.

Sheridan, M. F., and Wohletz, K. H., 1983, Hydrovolcanism: Basic considerations and review: Journal of Volcanology and Geothermal Research, v. 17, p. 1–29.

Wagner, G. A., Storzer, D., and Keller, J., 1976, Spaltspurendatierung quartarer Gesteinsgaser aus dem Mittelmeeraum: Neues Jahrbuchfür Mineralogie Monatschefte, p. 84–94.

MANUSCRIPT ACCEPTED BY THE SOCIETY MAY 5, 1986

Holocene rhyodacite eruptions on the flanks of South Sister volcano, Oregon

William E. Scott*
U.S. Geological Survey
Denver Federal Center
Denver, Colorado 80225

ABSTRACT

Almost 0.9 km^3 (dense-rock equivalent) of rhyodacite tephra, lava domes, and lava flows erupted from numerous vents on the southwest, southeast, and northeast flanks of South Sister volcano during late Holocene time. Eruptions occurred during two brief episodes between 2300 and 2000 ^{14}C yr B.P., separated by a dormant interval of as long as several centuries. The eruptions of each episode were probably fed by dikes, on the basis of the following: the alignment of vents, the chemical uniformity of eruptive products, and stratigraphic evidence that the eruptions of each episode occurred over a short interval of time.

Each eruptive episode began with the explosive eruption of air-fall tephra. Small pyroclastic flows and hot pyroclastic surges erupted from a few vents and traveled as far as 3 km. Rapid snowmelt accompanied the early phase of each episode and triggered small lahars. Each episode culminated with the extrusion of lava domes and flows.

The distribution of late Quaternary mafic vents around the area of Holocene rhyodacite vents suggests that a magma chamber with a maximum areal extent of 30 km^2 may lie beneath the south flank of South Sister. The chemical uniformity of the eruptive products of each episode is consistent with each having tapped a relatively small homogeneous portion of a compositionally zoned magma chamber of much greater volume than the erupted products. Alternately, if this chemical uniformity reflects the generation and rapid ascent and eruption of a crustal partial melt, then a large magma chamber need not be present.

INTRODUCTION

Rhyodacite[1] tephra, lava domes, and lava flows erupted during late Holocene time from more than 20 vents on the southwest, southeast, and northeast flanks of South Sister volcano in the central High Cascade Range of Oregon (Figs. 1 and 2). The area around South Sister has been the site of recurrent silicic volcanism during Pliocene and Quaternary time (Williams, 1944; Taylor, 1978, 1981; Mimura, 1984). Products of this activity vary in scale from voluminous ash-flow tuffs to small-volume tephra-fall deposits and lava domes. According to Williams (1944) and Taylor (1978), Holocene rhyodacite eruptions at South Sister began on the southwest flank at Rock Mesa and were followed by eruptions from aligned vents on the southeast and northeast flanks. Both authors recognized that explosive eruptions of tephra preceded the extrusion of lava flows and domes.

This report describes in greater detail the character and timing of these eruptions based largely on stratigraphic studies of fragmental deposits and lavas. In addition, evidence presented here and in Fink (1984) indicates that most of these eruptions were fed by dikes. The report concludes by speculating about the source of the Holocene rhyodacite magmas.

---

*Present address: Cascade Volcano Observatory, U.S. Geological Survey, 5400 MacArthur Boulevard, Vancouver, Washington 98661.
[1]The term rhyodacite is after Taylor's (1978) usage—rhyodacites have more than 68% SiO$_2$ and less than 4% K$_2$O.

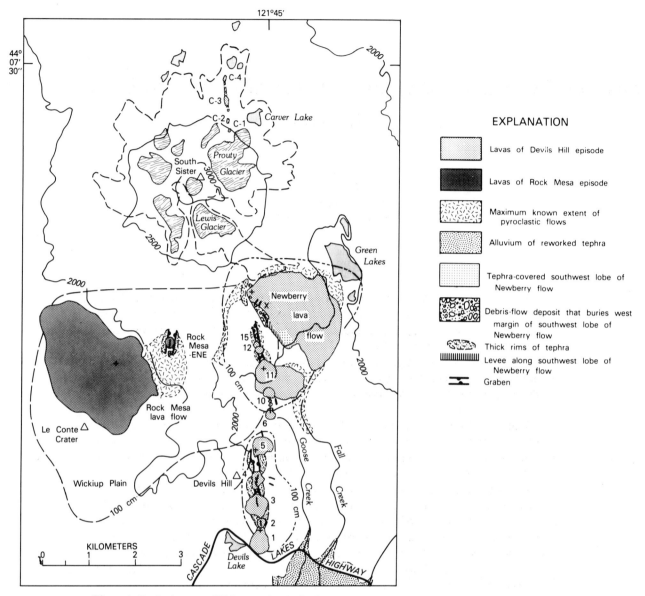

Figure 1. Geologic map of Holocene rhyodacite fragmental deposits and lavas on the flanks of South Sister volcano, Oregon. Lava vents along the Devils Hill chain are identified by numbers beginning with 1 at the south end; vents for small bodies are not numbered here. Lava vents along the Carver Lake chain are numbered C-1 to C-4. Heavy solid lines mark locations of tensional cracks, many of which occur over the vents of small domes and lie parallel to grabens. Crosses mark locations of vents for large lava flows. Small circles a short distance north and south of the Newberry vent are explosion craters. X marks area of fossil fumaroles on the southwest lobe of the Newberry lava flow. Long-dashed and dashed-and-dotted lines are 100-cm isopachs of the Rock Mesa and Devils Hill air-fall tephras, respectively. Short-dashed line shows maximum extent of glaciers during late Neoglacial time, which postdates the Devils Hill episode and ended late in the nineteenth century. Contours are in metres.

TYPES OF FRAGMENTAL DEPOSITS

The origin of fragmental deposits formed during the late Holocene eruptions is determined using standard field methods (e.g., Crandell and Mullineaux, 1975; Fisher and Schminke, 1984). The following key characteristics are used to identify the origin of fragmental deposits in the South Sister area:

Air-fall tephra forms plane-parallel-bedded, ground-mantling deposits. Beds are defined by variations in grain size, sorting, grading, and composition, which is based on proportion of white and gray pumice, dense to microvesicular, glassy rhyodacite, and accidental lithic fragments. Individual beds range from almost 100% pumice to almost 100% glassy rhyodacite; most clasts are angular to subangular. The content of accidental lithic fragments

Figure 2. Aerial photograph of the south flank of South Sister volcano showing Holocene rhyodacite lava flows and domes. R; Rock Mesa; RE, Rock Mesa-ENE; 1, vent 1; 5, vent 5; 6, vent 6; 11, vent 11; and N, Newberry vent of the Devils Hill chain of vents; D, Devils Hill. Talapus Butte, a basaltic scoria cone of latest Pleistocene age, is in the left foreground. Distance from 1 to N is about 5 km. View is toward the north-northwest. Photograph by C. Dan Miller, U.S. Geological Survey.

varies from less than 1 to 20%. Typically, distal lapilli beds are clast supported and well sorted, whereas near-vent tephra is more poorly sorted. This poor sorting is probably due to deposition from relatively low, pulsating eruption columns (Walker, 1981).

Pyroclastic-flow deposits form areally limited valley-filling or fan-shaped deposits of lapilli and blocks supported in an ash matrix. They formed from avalanches of hot pyroclastic debris and gases with overriding clouds of hot ash and gases. The deposits have typically 40–60 wt % ash (≤ 2 mm), of which about 20% is fine ash (≤ 0.0625 mm). Compared to Walker's (1971) field of pyroclastic-flow deposits on a plot of median grain size vs. deviation, median grain size of these South Sister pyroclastic-flow deposits is relatively coarse ($Md_\phi = 0$ to -1.5ϕ). Values for deviation ($\sigma_\phi = 2.6-4.0$) lie in the range that is common for pyroclastic-flow deposits. The deposits near South Sister are typically 30–200 cm thick, poorly sorted, and nonbedded; some are reversely graded and have a concentration of large pumice clasts in their upper parts. Most pumice clasts in the deposits are sub-round to round and some have pink rims or centers. Evidence that the deposits were emplaced at elevated temperatures includes their pink-colored tops that formed as a result of high-temperature oxidation and common bread-crusted and prismatically jointed clasts. Two pyroclastic-flow deposits sampled for paleomagnetic analysis display a single direction of thermoremanent magnetization of clasts, indicating emplacement at a high temperature (Hoblitt and Kellogg, 1979).

Hot pyroclastic-surge deposits are composed of ash and scattered rounded lapilli in plane-parallel and wavy beds and low-angle crossbeds. The deposits are typically interbedded with air-fall tephra and pyroclastic-flow deposits, and range from 20 to 150 cm thick. Some have a pale-pinkish-gray color, others are light to medium gray. They were deposited by dilute density currents of hot ash and gases derived from eruption columns or from the clouds overriding pyroclastic flows.

Lahars, or volcanic-debris-flow deposits, occur along a few valley floors in areas that received thick air-fall tephra. They are up to several meters thick and are composed of rounded pumice clasts and lithic fragments in a sandy matrix. The deposits are

TABLE 1. RADIOCARBON AGES OF ORGANIC MATERIALS ASSOCIATED WITH HOLOCENE RHYODACITE TEPHRAS AND CAYUSE CRATER SCORIA NEAR SOUTH SISTER, OREGON*

Lab Number	Age in ^{14}C year yr B.P. ± 1SD	Material dated and its stratigraphic position	Reference
Devils Hill tephra			
W-4013	1970 ± 200	Upper 1 cm of peat below Devils Hill tephra in Todd Lake meadow. Section D (Fig. 8) is similar to and close to dated section.	D.R. Mullineaux (1981, written commun.)
W-5208	2410 ± 80	Small charcoal fragments on weak soil formed in rock Mesa tephra and buried by Devils Hill tephra (Fig. 8, section F).	This report
W-5016	2480 ± 100	Small charcoal fragments from between 115-cm-thick Devils Hill and 20-cm-thick Rock Mesa tephra in roadcut on north side of Cascade Lakes Highway, 50 m west of Goose Creek.	This report
Rock Mesa tephra			
W-5556	2150 ± 150	Outer 1-2 mm of carbonized, 10-cm-diameter branch in Rock Mesa tephra (Fig. 8, section E).	This report
W-3402	2300 ± 200	Charcoal from center of 30-cm-diameter log in Rock Mesa tephra 500 m west of Rock Mesa.	Taylor (1978) and Meyer Rubin (1978, written commun. to D.R. Mullineaux)
W-4016	2560 ± 200	Upper 1 cm of peat below Rock Mesa tephra in Todd Lake meadow. Section D (Fig. 8) is similar to and close to dated section.	D.R. Mullineaux (1981, written commun.)
W-5021	2740 ± 70	Wood fragment on soil formed in Mazama ash and buried by Rock Mesa tephra, near section P (Fig. 8).	This report
Scoria of Cayuse Crater			
W-5209	9520 ± 100	1 to 2-cm-thick layer of orgnic-rich mud and sand overlying till and buried by 2.5 m of scoria of Cayuse Crater. Exposed in spring area 300 m SSE of Cayuse Crater. Sample could not be pre-treated, so is regarded as minimum.	This report

*Source: Meyer Rubin, U.S. Geological Survey (1984, written commun.)

poorly sorted and massive and many are clast supported. They formed from slurries of sediment and water probably generated when hot air-fall tephra melted snow on steep valley slopes.

CHARACTER AND AGE OF HOLOCENE RHYODACITE ERUPTIONS

Radiocarbon dates and stratigraphic evidence show that Holocene eruptions of rhyodacite occurred in two brief, closely spaced episodes between about 2000 and 2800 ^{14}C yr B.P. (B.P.; Table 1). Limited paleomagnetic data discussed later suggest that the eruptive episodes probably date from the younger part of this time interval. The combined volume of magma erupted during both episodes totals about 0.85 km^3. Almost 95% of this volume formed lava flows and domes; the remainder forms tephra and other fragmental deposits (Table 2).

The products of the two episodes are slightly different chemically; however, the tephras and lavas of each episode are virtually identical except for Na$_2$O content (Tables 3 and 4). This

TABLE 2. VOLUMES OF RHYODACITE (IN DENSE-ROCK EQUIVALENT*) ERUPTED FROM VENTS ON THE FLANKS OF SOUTH SISTER VOLCANO DURING ROCK MESA AND DEVILS HILL EPISODES

	Volume (km^3)		
Eruptive Episode	Tephra	Lava	Total
Rock Mesa	0.03	0.50	0.53
Devils Hill	0.02	0.30	0.32
	0.05	0.80	0.85

*Dense-rock equivalent = tephra volume x 0.3. Value of 0.3 is mean of estimates given by Walker (1981).

TABLE 3. MAJOR-ELEMENT ANALYSES OF LAVA AND BULK SAMPLES OF AIRFALL PUMICE OF HOLOCENE AGE, SOUTH SISTER VOLCANO, OREGON.*

	Pumice of Rock Mesa		Lava of Rock Mesa		Rock Mesa-ENE		Pumice of the Devils Hill chain of vents					
					Pumice	Lava						
	1	2	3	4	5	6	7	8	9	10	11	12
SiO_2	73.4	73.6	73.6	73.5	73.3	73.6	72.4	72.4	72.3	72.5	72.6	72.8
Al_2O_3	14.3	14.2	14.1	14.3	14.4	14.1	14.8	14.7	14.8	14.7	14.6	14.6
Fe_2O_3	0.63	0.62	0.61	0.61	0.62	0.61	0.71	0.72	0.72	0.70	0.70	0.67
FeO**	1.32	1.32	1.29	1.28	1.31	1.29	1.47	1.50	1.51	1.48	1.48	1.40
MgO	0.59	0.54	0.53	0.51	0.56	0.54	0.61	0.60	0.65	0.64	0.62	0.63
CaO	1.78	1.77	1.74	1.76	1.82	1.71	1.97	1.98	2.01	1.98	1.97	2.05
Na_2O	4.17	4.13	4.23	4.22	4.23	4.29	4.41	4.53	4.40	4.35	4.42	4.32
K_2O	3.40	3.41	3.41	3.42	3.36	3.44	3.15	3.07	3.11	3.18	3.16	3.15
TiO_2	0.28	0.28	0.27	0.28	0.28	0.27	0.32	0.33	0.33	0.31	0.32	0.30
P_2O_5	0.09	0.08	0.08	0.09	0.08	0.08	0.09	0.09	0.10	0.09	0.09	0.06
MnO	0.04	0.04	0.04	0.04	0.04	0.04	0.05	0.05	0.05	0.05	0.05	0.04
LOI	2.51	2.57	0.10	0.53	2.18	0.44	2.31	1.13	2.07	2.74	1.70	2.83
Sum	99.14	98.97	98.55	98.65	99.14	99.08	98.69	98.75	98.80	98.32	99.14	98.91

	Lava of the Devils Hill chain of vents												
	13	14	15	16	17	18	19	20	21	22	23	24	25
SiO_2	72.8	72.4	72.6	72.5	72.7	72.6	72.5	72.6	72.5	72.6	72.7	72.5	72.8
Al_2O_3	14.5	14.7	14.6	14.6	14.6	14.5	14.6	14.5	14.6	14.6	14.5	14.6	14.4
Fe_2O_3	0.67	0.70	0.69	0.72	0.68	0.70	0.68	0.68	0.69	0.68	0.71	0.70	0.69
FeO**	1.41	1.49	1.46	1.52	1.43	1.46	1.44	1.43	1.44	1.43	1.50	1.48	1.46
MgO	0.57	0.63	0.58	0.66	0.59	0.60	0.61	0.58	0.60	0.56	0.60	0.62	0.59
CaO	1.96	1.96	1.97	1.97	1.98	1.96	1.96	1.96	1.97	1.96	1.93	1.98	1.92
Na_2O	4.43	4.48	4.42	4.46	4.43	4.52	4.49	4.51	4.43	4.52	4.34	4.43	4.45
K_2O	3.20	3.15	3.18	3.15	3.18	3.18	3.18	3.19	3.20	3.19	3.20	3.19	3.21
TiO_2	0.31	0.33	0.32	0.33	0.32	0.31	0.32	0.31	0.32	0.32	0.31	0.32	0.31
P_2O_5	0.09	0.10	0.09	0.11	0.09	0.10	0.10	0.09	0.10	0.09	0.09	0.09	0.09
MnO	0.04	0.05	0.04	0.05	0.04	0.05	0.04	0.04	0.05	0.04	0.05	0.05	0.05
LOI	0.41	0.51	0.39	0.39	0.38	0.34	0.48	0.20	0.35	0.34	0.24	0.44	0.22
Sum	98.97	98.95	98.56	98.82	98.53	98.89	98.51	98.66	98.13	98.43	98.76	98.62	98.65

	Carver Lake chain of vents			
	26	27	28	29
SiO_2	72.6	72.7	72.8	72.7
Al_2O_3	14.5	14.5	14.5	14.5
Fe_2O_3	0.71	0.68	0.66	0.69
FeO**	1.49	1.41	1.40	1.44
MgO	0.60	0.53	0.52	0.53
CaO	2.02	2.03	1.99	2.01
Na_2O	4.50	4.49	4.52	4.50
K_2O	3.12	3.14	3.16	3.13
TiO_2	0.33	0.32	0.31	0.32
P_2O_5	0.08	0.08	0.08	0.08
MnO	0.05	0.04	0.04	0.04
LOI	1.50	0.50	0.34	0.68
Sum	99.85	99.67	100.03	100.00

*Analyses in weight percent, determined by X-ray fluorescence (J.A. Bartel, K. Stewart, and J. Taggart, analysts, U.S. Geological Survey, Denver). Analyses are recalculated to 100%. LOI = loss on ignition at 920°C for 1 hour. Sample locations are shown on Figure 10. Sums given are for the original analyses prior to calculations.

**Fe analyses from laboratory reported as Fe_2O_3. Fe_2O_3 and FeO values calculated assuming $Fe^{+2}/Fe_T = 0.7$, which is approximately the ratio reported by Clark (1983) for South Sister rhyodacite lavas.

TABLE 4. MEANS AND STANDARD DEVIATIONS OF ANALYSES (TABLE 3) OF ERUPTIVE PRODUCTS OF THE ROCK MESA, DEVILS HILL, AND CARVER LAKE VENTS
(See Notes below)

	Rock Mesa pumice and lava n=6	Devils Hill pumice n=6	Devils Hill lava n=13	Carver Lake pumice and lava n=4	Analytical error*	All Devils Hill and Carver Lake samples n=23	Analytical error†
SiO_2	73.5 ± 0.1	72.5 ± 0.2	72.6 ± 0.1	72.7 ± 0.1	± 0.13	72.6 ± 0.2	± 0.19
Al_2O_3	14.2 ± 0.1	14.7 ± 0.1	14.6 ± 0.1	14.5 ± 0	± 0.07	14.6 ± 0.1	± 0.05
Fe_2O_3§	0.62 ± 0.01	0.70 ± 0.02	0.69 ± 0.01	0.68 ± 0.02	± 0.03**	0.69 ± 0.02	± 0.02**
FeO§	1.30 ± 0.02	1.47 ± 0.02	1.46 ± 0.03	1.44 ± 0.04		1.46 ± 0.04	
MgO	0.55 ± 0.03	0.62 ± 0.02	0.60 ± 0.03	0.55 ± 0.04	± 0.02	0.60 ± 0.04	± 0.06
CaO	1.76 ± 0.04	1.99 ± 0.03	1.96 ± 0.02	2.02 ± 0.02	± 0.02	1.98 ± 0.03	± 0.02
Na_2O	4.21 ± 0.06	4.40 ± 0.07	4.54 ± 0.05	4.50 ± 0.01	± 0.09	4.45 ± 0.06	± 0.07
K_2O	3.41 ± 0.03	3.14 ± 0.04	3.18 ± 0.02	3.15 ± 0.03	± 0.01	3.16 ± 0.03	± 0.03
TiO_2	0.28 ± 0.01	0.32 ± 0.01	0.32 ± 0.01	0.32 ± 0.01	± 0.01	0.32 ± 0.01	± 0.01
P_2O_5	0.08 ± 0.01	0.09 ± 0.01	0.09 ± 0.01	0.08 ± 0	± 0.01	0.09 ± 0.01	± 0.01
MnO	0.04 ± 0	0.05 ± 0	0.05 ± 0.01	0.04 ± 0.01	± 0.01	0.05 ± 0.01	± 0.01
Ba	822 ± 34	793 ± 5	804 ± 6		± 40		
Be	1 ± 0	1 ± 0	1 ± 0		± 1		
Ce	38 ± 3	37 ± 1	41 ± 2		± 6		
Co	3 ± 1	4 ± 0	4 ± 0		± 4		
Cr	4 ± 1	3 ± 1	3 ± 1		± 4		
Cu	6 ± 1	8 ± 2	7 ± 1		± 7		
Ga	16 ± 1	15 ± 1	17 ± 1		± 2		
La	21 ± 1	21 ± 1	22 ± 0		± 2		
Li	32 ± 1	33 ± 4	32 ± 1		± 3		
Sc	3 ± 0	4 ± 1	4 ± 1		± 4		
Sr	193 ± 12	228 ± 4	223 ± 5		± 11		
V	15 ± 1	18 ± 1	18 ± 1		± 2		
Y	16 ± 1	16 ± 1	17 ± 0		± 2		
Yb	2 ± 0	2 ± 1	2 ± 0		± 2		
Zn	33 ± 2	39 ± 1	36 ± 2		± 4		

*Standard deviation of 49 samples prepared from the same standard; each sample counted once in same day.
†Standard deviation of one sample counted 62 times in a 6-month period.
§See second footnote in Table 3.
**Analytical error for total Fe reported as Fe_2O_3.

Notes: The groups of samples averaged in each of the first 4 columns were counted in separate lots; their standard deviations should be compared with the analytical errors in column 5. The analytical error that takes into account long-term drift (column 7) should be compared with the combined analyses of products of the Devils Hill and Carver Lake vents in column 6. Analytical errors for XRF analyses are from J. Taggart (U.S. Geological Survey, personal commun., 1985). Minor and trace elements, in parts per million, determined by induction-coupled plasma emission spectroscopy (P.H. Briggs and D.B. Hatfield, analysts, U.S. Geological Survey, Denver). Analytical errors for ICP analyses are from F. Lichte (U.S. Geological Survey, personal commun., 1985).

difference in Na_2O contents is probably due to hydration of the pumice (e.g., Lipman, 1965) as there is a strong inverse relation ($r^2 = 0.92$) between Na_2O content of the pumices and loss on ignition (Table 3). The products of the earlier episode are slightly more fractionated than those of the later episode, and contain about 1% more SiO_2 and about 0.3% more K_2O. The similar chemical composition of the products of both episodes is reflected in their mineralogic composition; all contain plagioclase, hypersthene, and very minor Fe-Ti oxides and hornblende.

The sequence of activity during each episode began with explosive eruptions of tephra from vents, most of which were aligned along fissures or grabens. During initial eruptions, tephra deposited on slopes of steep valleys on South Sister formed small lahars, probably the result of rapid snowmelt. Small pyroclastic flows and surges were produced at a few vents, generally as tephra eruptions waned. Each eruptive episode culminated with the extrusion of a lava dome or flow from some vents that had erupted tephra, whereas other tephra vents produced no lava.

During the first, or Rock Mesa, episode, eruptions occurred on the southwest and south flanks of South Sister from the vent of the Rock Mesa lava flow, as well as from a short linear vent system 1 km east-northeast of Rock Mesa, here called Rock Mesa-ENE (Fig. 1). Tephra from these vents was distributed mostly south and east to a maximum distance of about 30 km (Fig. 3).

The second, or Devils Hill, episode is named for a Pleistocene rhyodacite dome that lies just west of the southern part of a zone of aligned Holocene vents on the southeast flank of South Sister (Figs. 1 and 2). This zone consists of three segments to which Taylor (1978) gave separate names; however, in this report the entire zone is called the Devils Hill chain of vents. The vents that were sites of lava extrusions are identified by number, begin-

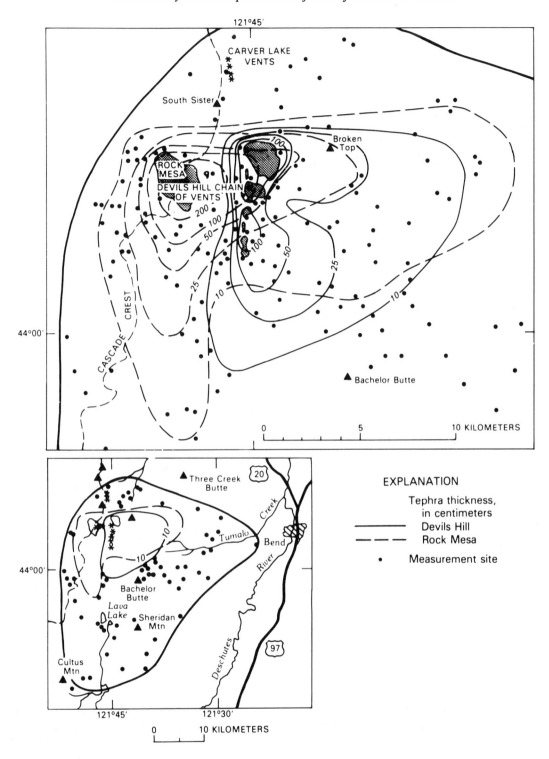

Figure 3. Isopachs of air-fall tephra erupted during the Rock Mesa and Devils Hill episodes. Where scale permits, lava flows and domes are outlined; others are located by asterisks. The heavy solid line delineates the approximate outer limit of tephra of both episodes except in the west and southwest, where only tephra of the Rock Mesa episode is present.

Figure 4. Tephra of the Devils Hill episode (DH) overlying tephra of the Rock Mesa episode (RM) exposed in a pit 3.5 km southeast of Broken Top. Knife is 20 cm long. The ash bed at the top of the Rock Mesa tephra was slightly reworked and weakly oxidized prior to the deposition of the Devil Hill tephra. Dark-colored unit below the Rock Mesa tephra is the A horizon of a soil formed in mixed Mazama ash and till.

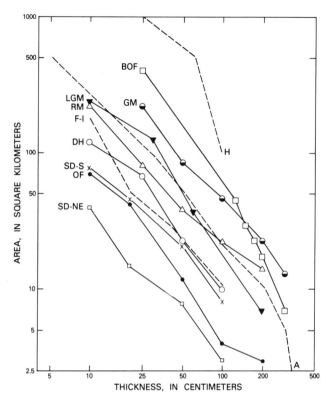

Figure 5. Log-log plot of isopach thickness against area enclosed by the isopach for air-fall tephras erupted during the Rock Mesa (RM) and Devils Hill (DH) episodes, as well as for other Holocene tephras of similar character in the western United States. These include the Glass Mt. (GM) and Little Glass Mt. (LGM) tephras at Medicine Lake volcano (Heiken, 1978); the tephra that preceded the eruption of the Big Obsidian Flow (BOF) in Newberry caldera (MacLeod and others, 1982); and the south lobe (SD-S) and northeast lobe (SD-NE) of the tephra erupted from the South Deadman vent, as well as the tephra erupted from the Obsidian Flow vent (OF) in the Inyo chain (Miller, 1985). Included for comparison are the tephras of eruptions of Hekla (H) in 1947 (Thorarinsson, 1967), Asama (A) in 1783 (Minikami, 1942), and Furnas (the upper part of prehistoric layer I (F-I); Booth and others, 1978). In Walker's (1973) eruption classification based on both dispersal and fragmentation, H and A lie in the Plinian field close to the arbitrary boundary with the sub-Plinian field, and F-I lies in the sub-Plinian field.

ning with 1 at the south end; however, the northernmost is termed the Newberry vent. Vent areas that produced only tephra are identified by the numbers of the adjacent lava vents. The many beds of tephra erupted from vents of the Devils Hill chain form a conformable sequence that was dispersed mostly east of the chain; this sequence typically overlies reworked and slightly weathered Rock Mesa tephra (Figs. 3 and 4).

The short linear vent system on the northeast flank of South Sister (C-1 to C-4 in Fig. 1; Wozniak, 1982), here called the Carver Lake vents, lies on strike with and was active at apparently the same time as the Devils Hill chain. Only a small amount of tephra erupted from the Carver Lake vents, and this was carried mainly eastward.

The pumiceous character of the rhyodacite air-fall tephra, its pattern of dispersal, and its associated pyroclastic flows and surges indicate the Plinian to sub-Plinian nature (Walker, 1973, 1981) of these eruptions. Figure 5 compares the dispersal of the Holocene Rock Mesa and Devils Hill tephras at South Sister with a few air-fall tephras of sub-Plinian and small, historic Plinian eruptions. Also plotted are other Holocene rhyodacite and rhyolite tephras in the western United States similar in character and extent to the South Sister tephras. Because (1) the two South Sister tephras plot between a subplinian and a small Plinian tephra and (2) each of the South Sister tephras comprise a sequence erupted from multiple vents, the eruptions at individual vents are probably best classified as sub-Plinian. The tephra erupted from individual vents likely had a dispersal pattern similar to that of tephra erupted from vents in the Inyo chain in California (SD and OF in Fig. 5; Miller, 1985), which have magma-equivalent volumes of about 0.01–0.04 km³.

Age of Eruptive Episodes

The available radiocarbon ages of charcoal, wood, and peat related to tephra that erupted from the Rock Mesa and Devils Hill vents (Table 1) and their calendar ages at the 95% confidence interval are shown in Figure 6. Although stratigraphic evidence presented below suggests that both episodes and an intervening dormant interval spanned a period of no more than several centuries, the radiocarbon ages range from sightly less than 2000 to almost 2800 B.P.; the range of calendar ages at the 95% confidence interval is even greater. The accuracy of the radiocarbon ages for dating the tephras may be limited by the following problems. First, the dated peat samples may be contaminated by younger organic matter because roots from the present surface vegetation extend through the peats. In addition, the charcoal and wood samples may come from the center of trees that were several centuries old at the time of the tephra falls. Considering these and other problems, the radiocarbon dates can provide only broad limits for the age of the eruptions. However, a date of 2150 ± 150 B.P. of a carbonized branch buried in Rock Mesa tephra should probably be considered one of the more reliable dates because the material dated is from the outer 1–2 mm of the branch, which was charred in growth position. This is supported by a date of 2300 ± 200 B.P. of the inner part of a carbonized log from the same deposit that was probably several centuries old at the time of the eruption, based on its diameter (Table 1).

Limited paleomagnetic information (Fig. 7) also suggests that the eruptive episodes occurred about 2300–2000 B.P. A pyroclastic-flow deposit of each of the two eruptive episodes was sampled and their directions of thermoremanent magnetization were determined using the techniques outlined by Hoblitt and others (1985). The paleomagnetic directions obtained from these deposits are plotted in Figure 7 along with paleomagnetic directions of other volcanic deposits in the northwestern United States that are dated between 3000 and 1800 B.P. (Champion, 1980). The data suggest an overall change of the magnetic field in a clockwise direction during this time period. The paleomagnetic directions of the Rock Mesa and Devils Hill deposits lie close to those of the youngest lavas erupted along the Great Rift in Idaho, which have mean dates of 2000–2200 B.P. (Kuntz and others, 1986). In contrast, directions of lavas in the central Oregon Cascades that are dated at about 2700–3000 B.P. have much shallower inclinations. A pyroclastic-flow deposit at Mount Rainier with an approximate age of 2350 B.P. has an intermediate direction. Thus, the ages of the Rock Mesa and Devils Hill episodes, based on the paleomagnetic data, are probably between 2300 and 2000 B.P.

The length of the apparent dormant interval between the Rock Mesa and Devils Hill eruptive episodes can be estimated

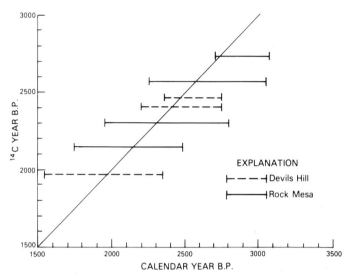

Figure 6. Radiocarbon dates (Table 1) on organic materials that immediately underlie or occur in the Rock Mesa and Devils Hill tephras plotted against their corresponding calendar ages at the 95% confidence interval using the tables of Klein and others (1982). The analytical errors of the ^{14}C dates are considered in these tables. The diagonal line corresponds to a 1:1 relation between ^{14}C and calendar ages.

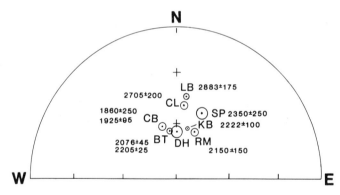

Figure 7. Equal-area, lower-hemisphere projection showing mean paleomagnetic directions and 95%-confidence ellipses for a pyroclastic-flow deposit of the Rock Mesa episode (RM; Fig. 8, sec. I) and a pyroclastic-flow deposit of the Devils Hill episode (DH; Fig. 8, sec. J). The declination, inclination, and α_{95} (95% confidence ellipse) of the points are RM-22.3°, 62.6°, and 2.0°, and DH-0.9°, 64.5°, and 3.2°. Other directions are from Champion (1980) and include the Clear Lake lava flow (CL) from Sand Mountain, Oregon; the Little Belknap lava flow (LB) at McKenzie Pass, Oregon; the South Puyallup block-and-ash flow (SP) at Mt. Rainier, Washington; the Kings Bowl lava flow (KB) at the south end of the Great Rift, Idaho; the Blue Dragon and Trench Mortar Flat lava flows (BT) at the Craters of the Moon, Great Rift, Idaho; and the Cave basalt lava flow (CB) at Mount St. Helens, Washington. The ^{14}C date of the Rock Mesa episode is from this report; the dates of the Great Rift lavas, which are the means of several dates, are from Kuntz and others (1986); and the other dates are from references listed in Champion (1980). No attempt was made to correct the data for differences in magnetic-field direction due to geographic location, because the present magnetic field directions in these areas vary by an amount about equal to or less than the uncertainties represented by the confidence ellipses.

from the degree of development of buried soils (Fig. 4) and from the thickness of peat between the tephras. The character of the contact between the tephras is variable. At most localities, a weakly developed soil had formed in a zone of reworked Rock Mesa tephra prior to burial by Devils Hill tephra; however, in other places, such as some roadcuts along the Cascade Lakes Highway, the Devils Hill tephra overlies little-altered Rock Mesa tephra. The following discussion focuses on the better-developed buried soils.

Soils formed in Rock Mesa tephra and subsequently buried by Devils Hill tephra are much more weakly developed than soils formed in Devils Hill tephra in the past ca. 2000 yr. The buried soils consist of one or more of the following horizons that overlie a weakly oxidized C horizon from 5 to 15 cm thick: O horizons, which are surface accumulations of partly decomposed organic matter, are only locally present; A horizons, which consist of mineral soil enriched in humified organic matter, are generally weak and thin (≤ 3 cm); E horizons, which consist of light-colored mineral soil relatively leached of iron compounds and formed in this area under mountain hemlock forests, are thin (≤ 2 cm) and very discontinuous. In contrast, the soil profile developed in Devils Hill tephra under mountain hemlock forest during the past ca. 2000 yr typically has (1) an O horizon at least 2 cm thick; (2) a 2- to 8-cm-thick continuous A horizon that is much more organic rich than the A horizon of the buried soil described above; (3) a 2- to 5-cm-thick E horizon; and (4) a 10- to 15-cm thick color B and (or) oxidized C horizon that is more intensely oxidized than the oxidized C horizon of the buried soil. This marked difference in profile development suggests that the dormant interval between the eruptive episodes was certainly less than 1000 yr but probably at least a few hundred years.

As much as several centimeters of peat or meadow sediments that occur between the tephras (Fig. 8, sections D, H) also provide evidence of the length of the dormant interval. By using the local rate of accumulation of the post-Devils Hill (post-2000 yr) sediments, the approximate time needed for accumulation of the interbedded sediments ranges from 100 to 500 yr.

In light of the preceding evidence, the length of the dormant interval between the eruptive episodes is unlikely to have been longer than a few hundred years, but is probably at least 100 yr.

Rock Mesa Episode

During the Rock Mesa episode, about 0.5 km^3, or more than one-half the total volume of Holocene rhyodacite (Table 2), erupted from the Rock Mesa and Rock Mesa-ENE vents. Initial tephra eruptions accompanied locally by lahars and pyroclastic flows and surges preceded the emplacement of lava flows and domes. A dome on the highest part of the Rock Mesa lava flow is presumed to mark part of the buried Rock Mesa vent area; the Rock Mesa-ENE vent is a well-defined N10°E-trending graben that contains several domes and has thick (>10 m) accumulations of tephra along its margins.

Air-fall tephra. Tephra of the Rock Mesa episode was distributed by winds mainly in two lobes, one south and the other east of the vents (Fig. 3). Although the prevailing winds at altitudes that would be reached by sub-Plinian and small Plinian eruption columns blow dominantly from westerly directions, winds from the north and north-northeast that would disperse tephra southward presently occur at least 5–10% of the time during all months except July and August (data from "Winds Aloft Summaries" of the Air Weather Service, U.S. Air Force, obtained from the U.S. National Climate Center at Asheville, North Carolina). Typically, Rock Mesa tephra overlies soils formed in 6800-yr-old Mazama ash and (or) till of latest Pleistocene age. The tephra is coarse grained and more than 10 m thick near vents and extends downwind as far as 30 km. In these distal areas, it consists of medium to coarse ash and scattered fine lapilli mixed into the upper few centimetres of soils. The farthest distance from vents that well-preserved layers are found is about 20 km, where 1- to 3-cm-thick layers of ash and fine lapilli occur in meadows and bogs.

The tephra of both lobes, as well as the lava extrusions in both vent areas, have the same chemical composition (Table 3) and mineralogy, with phenocrysts of plagioclase, hypersthene, and very minor Fe-Ti oxides and hornblende. Because of these similarities, the proximity of the vents, the lack of exposures, and the difficulty in tracing and correlating beds, it is not possible to determine the source vent for individual beds or bed sets except in a few cases. One useful marker is a brick-red, possibly phreatomagmatic, ash (Fig. 8, bed R) that typically forms a thin (<1–5 cm) bed, except near its Rock Mesa-ENE source vent, where it forms two 25-cm-thick beds of lapilli and ash separated by gray ash and lapilli. This marker occurs in the upper part of much of the east lobe and also overlies tephra of the northeast part of the south lobe (Fig. 8, section B).

The Rock Mesa tephra is characterized by multiple beds of lapilli and ash that have contrasting thickness, grain size, sorting, and grading. The lack of evidence of subaerial exposure within the tephra sequence, such as burrowing by animals, growth of vegetation, and reworking or weathering (Fig. 8), suggests that it was deposited over a short period of time, probably less than 1 yr. Furthermore, contacts between many beds are gradational, indicating that the eruption did not cease in the time between their deposition. Rather, the vigor of the erupting column varied and (or) the wind velocity or direction changed during the eruption. These relations suggest that the tephra sequence was probably deposited during a single, nearly continuous eruptive episode.

At altitudes near 2000 m, the lower part of the tephra sequence locally contains disrupted beds, whereas the upper part is regularly bedded (Fig. 8, section E). In some of these places, faults with displacements of as much as several tens of centimeters cut the lower part to almost the entire tephra sequence. The faults do not displace the pre-tephra surface, but die out near the base of the deposit in tephra that has highly disrupted bedding or in tephra that has been so thoroughly mixed that bedding is not visible. These features and their isolated occurrences are thought to reflect the meltout of snowbanks and the consequent collapse

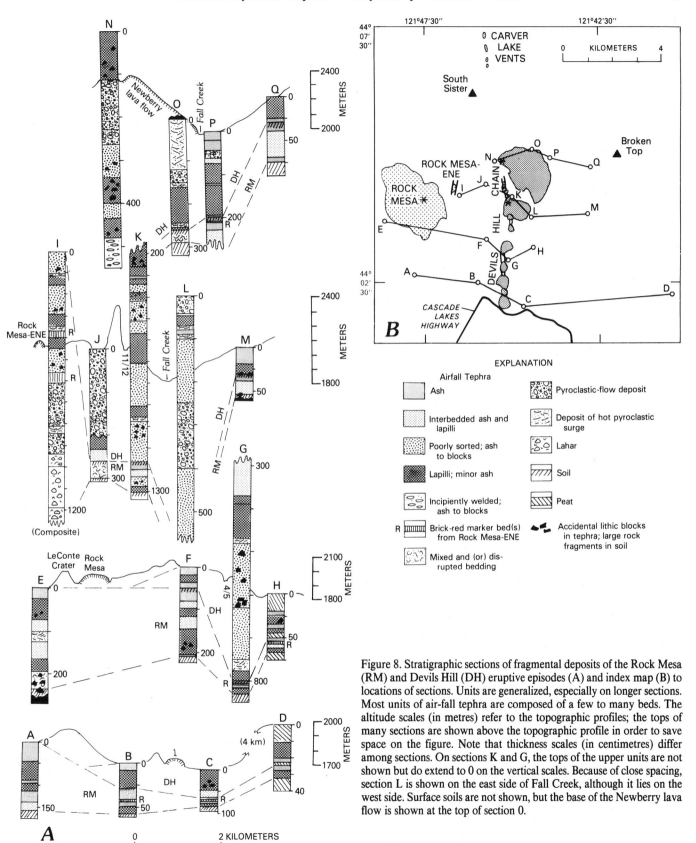

Figure 8. Stratigraphic sections of fragmental deposits of the Rock Mesa (RM) and Devils Hill (DH) eruptive episodes (A) and index map (B) to locations of sections. Units are generalized, especially on longer sections. Most units of air-fall tephra are composed of a few to many beds. The altitude scales (in metres) refer to the topographic profiles; the tops of many sections are shown above the topographic profile in order to save space on the figure. Note that thickness scales (in centimetres) differ among sections. On sections K and G, the tops of the upper units are not shown but do extend to 0 on the vertical scales. Because of close spacing, section L is shown on the east side of Fall Creek, although it lies on the west side. Surface soils are not shown, but the base of the Newberry lava flow is shown at the top of section O.

of tephra mantles either during or following deposition of the tephra. The isolated distribution of these features suggests that the eruptions occurred during early to middle summer, which is the time today during which isolated snowbanks are present at these altitudes.

Lahars. The early tephra eruptions also caused melting of snow on the south slope of South Sister, which resulted in the formation of pumiceous lahars. The deposits of these lahars are exposed at the base of some tephra sections along gullies near the Rock Mesa-ENE vent (Fig. 8, section I). These lahars affected areas mostly along valleys within about 1 km of the vent.

Pyroclastic flows and surges. Several pyroclastic flows and hot pyroclastic surges erupted from the Rock Mesa-ENE vent, whereas only surge deposits have been found around the Rock Mesa vent (Fig. 8, sections E, I). Gullies a few hundred meters east of Rock Mesa-ENE expose several pyroclastic-flow deposits that can be traced about 1 km from the vent (Fig. 1). They may have traveled as far as the low-gradient surface upon which the Rock Mesa flow was later extruded; however, lack of exposures and burial by thick alluvium of reworked tephra hinder further tracing. The pyroclastic-flow deposits are typically 1–1.5 m thick and poorly sorted and are composed of rounded lapilli of pumice and microvesicular rhyodacite in an ash matrix. Their relatively coarse grain size (Md_ϕ = 0 to -1.5ϕ), small volume, and restricted distribution indicate that the pyroclastic flows that produced them were probably of low mobility. Deposits of hot pyroclastic surges are restricted to within about 2 km of the vents and consist of several tens of centimeters of light-gray to pinkish-gray ash with parallel and wavy bedding and crossbedding.

Lava flows and domes. Following the tephra eruptions, a lava flow with a volume of about 0.5 km^3 was extruded from the Rock Mesa vent. The highest point on the flow is a dome that is assumed to have been formed over the vent by the last-erupted lava (Fig. 1). The center of the graben at the Rock Mesa-ENE vent was partly filled by a 200-m-diameter dome that has several short radial fractures and a central furrow oriented parallel to the graben. A few small (tens of meters in diameter) domes in the south end of the graben have partial mantles of tephra and may have been emplaced prior to the end of the tephra eruptions. Alternately, these small domes may have carried the mantles upward from the tephra-covered floor of the graben, because the mantles are discontinuous and the higher parts of the domes are barren of tephra. In this case, the small domes would postdate the tephra eruptions and may be coeval with the 200-m dome.

Devils Hill Episode

During the Devils Hill episode, tephra and lava domes and flows erupted from the north-trending, 5-km-long Devils Hill chain of vents (Fig. 2). This chain is composed of several en echelon segments with orientations of N3°W to N13°W. At apparently the same time, the 1.2-km-long Carver Lake chain of vents, which trends N6°W and is on strike with the Devils Hill chain, erupted on the northeast flank of South Sister (Fig. 1). Tephra from the Devils Hill chain overlies much of the east lobe of Rock Mesa tephra and the northeast part of the south lobe (Figs. 3 and 8).

Many vents along the Devils Hill chain lie in grabens with steep bounding faults in bedrock and till. Analogous to grabens formed above upwardly propagating dikes in basaltic volcanoes (Pollard and others, 1983), these grabens probably formed immediately prior to the beginning of pyroclastic eruptions. Normal faults locally cut tephra on the margins of the grabens; however, their discontinuity and small displacement suggest they may be related to slumping of tephra into the graben rather than to displacements on the graben faults. Faulting had ceased prior to lava eruptions, as the graben faults are locally buried by lava domes and flows. Evidently, activity began with graben formation followed by the opening of tephra vents along most of their length. The occurrence of thick rims of tephra along parts of the chain indicates that as the eruption progressed, activity focused at certain vents, which became principal conduits. A similar sequence of events has been observed during fissure eruptions in Hawaii and is due to preferential enlargement of some vents along the dike by erosion during the eruption (Delaney and Pollard, 1981). A similar process probably operates in silicic dikes. This is suggested by thickening below a principal vent of the dike that fed the Inyo chain of domes in eastern California (Eichelberger and others, 1985). The Devils Hill episode culminated with the extrusion of lava flows or domes from about 16 vents along the chain. These range from small domes less than 100 m in diameter to the 0.2-km^3 Newberry lava flow, which accounts for about 60% of the total volume erupted during the Devils Hill episode (Table 2).

Air-fall tephra and sequence of eruptions along the chain. Stratigraphic relations of the Devils Hill tephra suggest that pyroclastic eruptions occurred along the entire length of the chain of vents without significant breaks in time. As in the Rock Mesa tephra, near-vent tephra sequences contain multiple beds of contrasting grain size, thickness, grading, and sorting. Common gradational contacts between beds reflect essentially continuous deposition as changes occurred in the height and location of eruption column(s) and (or) in the wind velocity and direction. No evidence of reworking or weathering has been found within these near-vent sequences. Similar relations are found in distal areas east of the chain of vents, where tephra was probably derived from several vents along the chain. Here, the Devils Hill tephra consists of one (Fig. 4) to a few conformable, graded beds (Fig. 8, section H).

Because of the similar composition of individual beds or bed sets, they cannot be correlated over large areas, and the timing of eruptions along the chain cannot be determined with certainty. However, the lithology of accidental fragments help locally to limit the tephra source to certain sections of the chain. For instance, northern vents of the chain erupted a conspicuous proportion of reddish and purplish vesicular clasts of andesite and dacite(?), whereas southern vents erupted very little material of this color. In the area that lies between the north and south lobes of Devils Hill tephra (Fig. 3), beds containing these clasts are conformably overlain by beds containing few or none of these clasts, suggesting that at least some pyroclastic eruptions at the northern end of the chain preceded shortly those from the southern end.

The tephra from the Carver Lake vents on the northeast flank of South Sister forms a much thinner and less extensive blanket than tephra from vents along the Devils Hill chain. Isopachs of the tephra deposits near and east of the Carver Lake vents are not shown in Figure 3 for several reasons. First, thick near-vent accumulations of tephra were largely removed by erosion during late Neoglacial time, as were parts of the domes. Also, beyond the limit of late Neoglacial ice, thin tephra and conspicuous bread-crusted bombs and lapilli of gray pumice are present, but have largely been reworked by slope processes. The original thickness of tephra in this area is not known, but probably was less than 25 cm. Finally, at lower altitudes farther east in meadows and forested areas, the tephra is less than 10 cm thick. Typically the tephra is ≤5 cm thick, is mostly gray ash, and contains scattered, gray bread-crusted lapilli as large as 5 cm. The thinness, restricted areal distribution, and fine grain size of these deposits, when compared with tephra erupted from the other Holocene vents, indicate that the pyroclastic eruptions that preceded the emplacement of the Carver Lake domes were relatively brief and small.

The tephra from the Carver Lake vents overlies slightly weathered Rock Mesa tephra or thin sediments that overlie Rock Mesa tephra; however, its stratigraphic relation to Devils Hill tephra is less certain because of the mineralogical similarity between the Devils Hill and Carver Lake eruptive products. Both contain phenocrysts of plagioclase and hypersthene and very minor Fe-Ti oxides and hornblende. In the area between Carver Lake and the north flank of Broken Top, limited evidence suggests the tephra from the Carver Lake vents probably conformably underlies the Devils Hill tephra; however, it is possible that the tephras from the Carver Lake and Devils Hill vents are partly interbedded. In either case, the tephras appear virtually coeval. This is also consistent with the identical chemical composition of the Devils Hill and Carver Lake products (Tables 3 and 4) and the possibility that both eruptions were fed from the same dike system, as is discussed in a later section of this report.

Lahars. Pumice-rich lahars are present in the Fall Creek valley downstream from the Newberry lava flow. Their stratigraphic position is uncertain, but is likely below section L (Fig. 8). Other, largely lithic debris-flow deposits of probable nonvolcanic origin postdate the incision of Fall Creek into the Devils Hill deposits. Goose, Fall, and Soda creeks reworked Devils Hill deposits to form a large alluvial fan at the head of Sparks Lake basin, which lies in the southeastern part of Figure 1.

Pyroclastic flows and surges. Pyroclastic flows and surges erupted from the Newberry vent—which lies at the northernmost and highest part of the Devils Hill chain— and perhaps from near vent 11 as well. These traveled east into the Fall Creek valley and into the unnamed valley west of the chain (Figs. 1 and 8). Deposits of the pyroclastic flows occur at or near the upper part of the tephra sequence, which suggests they may have originated by gravitational collapse of the eruptive columns that dispersed the tephra (e.g., Sparks and Wilson, 1976). The deposit west of the chain (Fig. 8, section J) contains a large proportion of pink, rounded pumice lapilli and small blocks, and has a pronounced pinkish-gray upper part. The deposit exposed below the base of the Newberry lava flow east of the chain is overlain by a 1–1.5-m-thick surge deposit of fine to coarse ash (section O). The pyroclastic-flow and surge deposits can be traced about 3 km from their source vent.

Lava flows and domes. Lava flows and domes erupted along the Devils Hill and Carver Lake chains after most pyroclastic activity had ceased; this is indicated by the absence of pyroclastic deposits on the surface of all the flows and domes, except for part of the southwest lobe of the Newberry flow (Figs. 1 and 9). This lobe, which overlies thick tephra erupted from vents between vent 11 and the Newberry vent, is partly covered with several tens of centimeters of angular ash and lapilli of microvesicular to dense, glassy rhyodacite. This tephra is the youngest bed erupted from vents 11–15 and extends less than 1 km from these vents. Its limited distribution is thought to reflect the feeble eruption of mostly degassed magma immediately prior to the emplacement of the lava flow and domes at vents 11–15. Therefore, the southwest lobe of the Newberry flow was emplaced before or perhaps during the final stage of pyroclastic activity at vents 11–15. In addition, the continuity of the flow ridges on the distal part of the southwest lobe of the Newberry flow with the flow ridges on the distal part of the main lobe to the northeast indicates that the distal parts of both lobes are probably coeval. The northeast lobe does not have the tephra mantle probably because it lay beyond the tephra's fallout area.

In addition to a partial mantle of tephra, the southwest lobe of the Newberry flow has several unusual features not found in other Holocene rhyodacite lava flows and domes in the area. These features occur along the southwest margin of the lobe; the northeast margin is typical of the other flows in displaying well-defined flow ridges and steep, talus-covered margins. The most atypical feature is a levee along the southwest margin of the lobe that is composed of blocks that resemble basal flow breccia (Fig. 9). These blocks are composed of pebble- to boulder-size fragments of black obsidian and gray, microvesicular rhyodacite that are set in a weakly to firmly indurated matrix of light reddish-brown, pulverized rhyodacite. Evidence of shearing is present in many blocks. In order to form a levee of blocks of breccia apparently derived from the base of the flow, the flow must have been greatly disrupted, because the material in levees is derived mainly from the top of a flow. Also atypical are several deep, open, tension cracks that trend normal to the steep (25°) slope of the lobe. These cracks occur along the southwest margin of the lobe nearest to the Newberry vent. In addition, the surface of the lobe downslope from the cracks has numerous fossil fumaroles, which are not found on any of the other lava flows. The fumaroles are marked by areas stained red, white, and yellowish-brown and

Figure 9. Photograph of the southwest margin of the Newberry lava flow showing some of the features that are described in the text. View is to the southeast. Trees are growing on the southwest lobe of the flow, which is partly covered by tephra. The levee (LV) ranges from 20-50 m wide and is composed of blocks that include basal flow breccia (see text). Younger debris-flow and alluvial deposits (DA) bury part of the lobe. The barren, northeast lobe of the flow is in the left background.

indicate there was a source of water below the southwest margin of the lobe, but on this steep slope it is unlikely that the source was a stream or pond.

One explanation for these features is that the southwest part of the southwest lobe flowed onto a steep, snow-covered slope. Melting of the snow caused the southwest margin to collapse and initiated sliding, which was facilitated by the presence of meltwater between the flow and an underlying layer of snow or the ground. This collapsing and sliding disrupted the flow so that blocks of basal breccia were brought to the surface. These blocks were then in a position to be incorporated in a levee as the disrupted part of the flow slid farther downslope. The opening of the tension cracks was probably also related to this sliding. Steam and water vapor derived from the melting snow escaped upward and formed the fumaroles. Some of the debris-flow deposits that bury the upslope part of the southwest lobe may also be related to rapid snowmelt during and immediately following extrusion of the lobe.

DISCUSSION

Evidence for Dikes Feeding the Eruptions

Fink and Pollard (1983) presented several types of evidence that linear zones of vents such as the Devils Hill chain were fed by a single dike or segments of a single dike. This evidence includes alignment of vents, paired zones of cracks and grabens between vents, and linear patterns of fractures and other structures on domes that parallel the alignment. These features, along with the chemical similarity of the erupted products from aligned vents (Bailey and others, 1983), and the apparent synchroneity of their eruptions, were used by Miller (1985) to hypothesize that the vents along the Inyo volcanic chain were fed by a dike. This has now been confirmed by core drilling that intersected a dike between the vents of two of the Inyo lava flows (Eichelberger and others, 1985).

The Devils Hill chain of vents, the Carver Lake vents, and the Rock Mesa-ENE vent display firm structural evidence that they were fed by dikes; structural characteristics of the Rock Mesa vent area are unknown because it is completely covered by the Rock Mesa lava flow. Aligned vents, grabens, ground cracks, and linear fractures over the vents of lava domes and flows parallel to vent alignments are shown in Figure 1. In addition, the chemical (Tables 3 and 4) and mineralogic uniformity of the tephras and lavas erupted during each episode is consistent with dike-fed eruptions, because it strongly indicates that each eruption was fed by magma from a single source. In contrast, tapping of a magma chamber by many pipelike conduits might be expected to sample magma from several different parts of the chamber and therefore vary somewhat in chemical composition.

The apparent near synchroneity of eruptions from several to many vents during each eruptive episode also provides evidence that the eruptions were fed by dikes. The stratigraphic relations, described previously, that suggested the eruptions occurred over a short interval of time include: (1) tephra of the Rock Mesa and Devils Hill episodes each form single depositional sequences with no marked disconformities or other evidence of significant interruptions during their deposition, and (2) tephra eruptions during each episode had largely ceased by the time lava domes and flows emerged, because—except for the southwest lobe of the Newberry flow—none of the lavas are overlain by tephra. More complex stratigraphic relationships would seem likely if tephra and lava vents had been active sporadically along the chain over an extended period of time.

During both eruptive episodes, the feeder dikes were locally enlarged, probably by brecciation and erosion (e.g., Delaney and Pollard, 1981), and these areas emerged as principal conduits. Therefore, the dikes did not supply magma uniformly to vents, especially during the lava-extrusion stage of each episode. Figure 11 is a vertical section through the Devils Hill and Carver Lake vents showing the position of the inferred feeder dike for both vent systems. For the Devils Hill chain, the cumulative-volume curve shows a direct relation between altitude and volume erupted; about 60% of the total emerged from the 2350-m-high Newberry vent, and much smaller volumes erupted at lower altitudes to the south. To the north, the Carver Lake vents, which lie at 2300–2500 m, erupted less than 1% of the total. The lack of vents along the chain above an altitude of 2500 m (unless there are some concealed by Prouty Glacier) may be due to the large volume of lava erupted from the Newberry vent, which would have acted to reduce magmatic pressure in the dike and to retard further upward propagation. As mentioned earlier, limited evidence suggests the tephra eruptions at the Carver Lake vents may have shortly preceded those along the northern part of the Devils Hill chain. However, if lava extrusion in both areas was concurrent, then the large volume extruded from the Newberry vent may have reduced the supply of magma to the northern end of the dike and resulted in the eruption of only a small volume of lava. Another possibility is that the dike was symmetric about the Newberry vent and, consequently, vents to the south erupted more lava than those to the north because they lay at considerably lower altitudes.

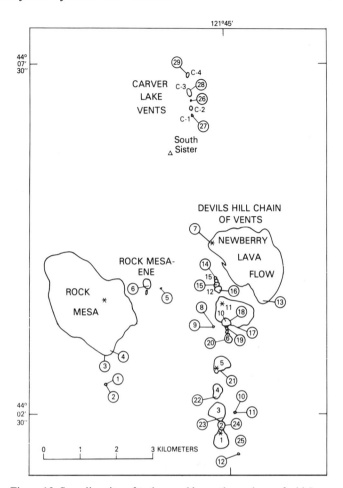

Figure 10. Sampling sites of tephras and lavas, the analyses of which are given in Table 3. Circled numbers correspond to sample numbers in Table 3. Some tephra sites have two numbers. Sample 1 was taken from a 20-cm depth and sample 2 from a 70-cm depth in a sequence of Rock Mesa tephra 160 cm thick. Sample 8 is gray, microvesicular pumice from the base of a 6-m section of tephra from vents 6–11 of the Devils Hill chain; sample 9 is white pumice from a 4-m depth in the same sequence. Sample 10 is white pumice and sample 11 is gray microvesicular pumice from the same bed in the base of a 3-m-thick sequence of tephra from the vent 3–4 area of the Devils Hill chain. Numbers that are not circled identify selected Devils Hill and Carver Lake vents.

Speculations about the Source of Holocene Rhyodacite

The evidence discussed above indicates strongly that the Holocene flank eruptions of rhyodacite were fed by dikes, and the geometry of the dikes inferred from vent patterns suggests two alternatives for the magma's source. One possibility, adapted from Nakamura (1977), is that the magma for the flank eruptions was supplied by a radial dike propagating outward from the central conduit of South Sister. Presumably, the central conduit would have tapped a magma chamber lying beneath the volcano. Because South Sister lies in a field of north-south compressive stress (Fig. 12), the trend of the dike should parallel this direction, except near the central conduit, where it should have a more

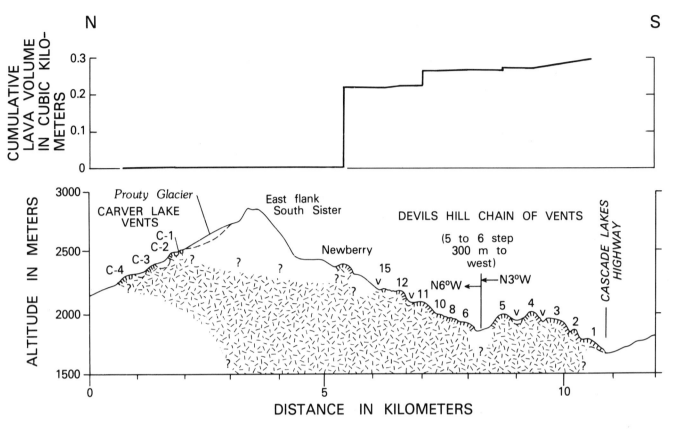

Figure 11. Vertical section along the trend of the Devils Hill chain of vents (1 to Newberry) and Carver Lake vents (C-1 to C-4) showing inferred extent of feeder dike (patterned; lower sketch) and the cumulative volume of lava erupted from vents or groups of vents (upper graph). The vents from 6 to C-4 are projected onto a section oriented N6°W, even though the strike of individual segments ranges up to N13°W. V's show locations of grabens not filled by lavas. Queries indicate great uncertainty in estimating the extent of the inferred dike. The volume erupted from each vent is shown by a vertical line, except where small volumes from closely spaced vents are represented by a gently sloping line (as for the 12 to 15 segment). Tephra volume is not included, as it represents only about 5% of the total volume of erupted magma. Based on the tephra's distribution (Fig. 3), subequal amounts erupted from the 1–5 and the 6-Newberry vent areas, and an insignificant volume erupted from the C-1 to C-4 vent area.

radial orientation. The Rock Mesa-ENE vent, which lies close to the central conduit, has a radial trend. Also, the segments of the Devils Hill chain follow this pattern. The north segment (vent 6 to Newberry vent), which is closer to the conduit of South Sister, has a more radial trend (N9°W to N13°W) than the south segment (vents 1 to 6), which has a N3°W trend (Figs. 1 and 2). In this scenario, the rhyodacite magma rose initially through the central conduit of South Sister and then migrated outward along a radial dike to emerge at the Rock Mesa and Rock Mesa-ENE vents. Because these two vent systems are not on strike, they presumably represent two segments of a dike that were rotated different amounts as they approached the surface. Several centuries later, slightly less-fractionated magma fed a dike that propagated south and north from the conduit to emerge at the Devils Hill and Carver Lake vents; the near-conduit segment from vent 6 to the Newberry vent had a more radial trend than the south segment. In conflict with this view, the Carver Lake vents, which lie closer to the conduit than any other vents, show no tendency for a radial orientation.

Alternately, the alignment of the Carver Lake and Devils Hill vents may reflect the emplacement of a segmented dike parallel to the direction of maximum horizontal compressive stress. South Sister effected very little radial control on the orientation of the dike except to cause slight rotation of the north segment of the Devils Hill chain of vents. Again, the nonradial orientation of the Carver Lake vents is puzzling. In this view, the central conduit of the volcano was probably not utilized in the magma's ascent, which is supported by the apparent lack of concurrent eruptive activity at the summit of South Sister. In addition, because more than 90% of the Holocene rhyodacite was erupted from the Rock Mesa vent, vent 11, and the Newberry vent—all of which lie close to the south flank of the main cone—the magma chamber that fed the eruptions may underlie the south flank, as Clark (1983) and Bacon (1985) have suggested.

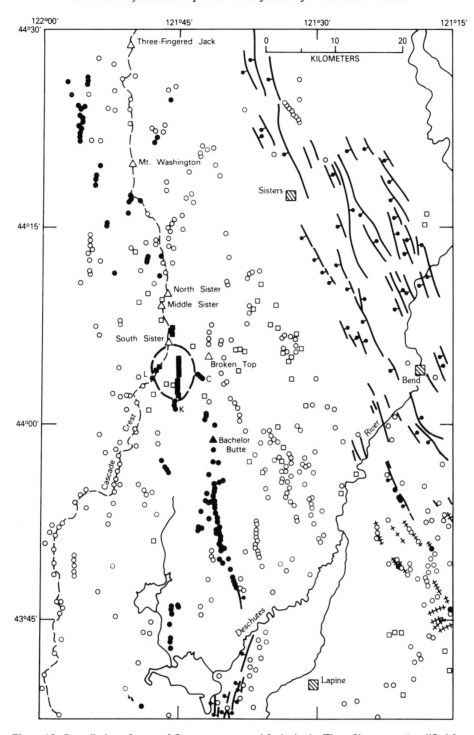

Figure 12. Compilation of vents of Quaternary age and faults in the Three Sisters area (modified from Peterson and others, 1976; MacLeod and others, 1982; and Taylor and MacLeod [written commun.] *in* Bacon, 1985). Vent alignments reflect the north-south direction of the maximum horizontal compressive stress (see discussion and references in Zoback and Zoback, 1980). Triangles are large stratovolcanoes; circles are vents of mafic (basalt and basaltic andesite), mostly monogenetic volcanoes; squares are vents of silicic (dacite and rhyodacite) lava domes and flows. Numerous silicic vents high on the south flank of South Sister are not shown. Solid symbols represent vents of latest Quaternary (<15,000 yr B.P.) age; open symbols are vents of pre-latest Quaternary age. L = Le Conte Crater; K = Katsuk and Talapus buttes; C = Cayuse Crater. Heavy solid lines are faults; the bar and ball symbol is on the downthrown side. Crossed lines are fissures. The dashed circle represents the maximum extent of a hypothesized magma chamber south of South Sister (see text).

Although little is known about the size of the magma chamber that supplied the nearly 1 km^3 of late Holocene rhyodacite, some constraints are apparent. A maximum areal limit for a silicic magma chamber near South Sister is suggested by the 30-km^2 area that encloses the Rock Mesa and Rock Mesa-ENE vents plus the Devils Hill chain of vents, and in which no vents for mafic magmas have been found (Fig. 12). The lack of eruptions of mafic lavas in this area may indicate the presence of silicic magma that has prevented the rise of relatively more dense mafic magma. In contrast, mafic magmas have erupted during latest Pleistocene and Holocene time on all sides of this 30-km^2 area, as well as for considerable distances to the north and south. On the west, Le Conte Crater erupted basaltic andesite during postglacial time, but prior to 6800 B.P., as it is unglaciated but covered by Mazama ash. On the basis of petrographic and chemical evidence, Clark (1983) suggested that the Le Conte magma interacted with both rocks and magma of silicic composition during its ascent. On the south, a north-south alignment of vents, including Talapus and Katsuk Buttes, erupted basalt in contact with receding glaciers of latest Pleistocene age. On the east, Cayuse Crater and several small vents to the northwest erupted basalt in early Holocene time; organic matter that underlies the scoria from Cayuse Crater has a radiocarbon age of 9500 B.P. (Table 1). Wozniak and Taylor (1981) suggested the basaltic andesite summit cone of South Sister to the north, and its well-preserved crater was formed in latest Pleistocene or Holocene time. The lack of fragmental deposits around the base of South Sister associated with the eruption of the summit cone, however, suggests to me a late glacial age, when ice still surrounded South Sister.

If this 30-km^2 area does indeed approximate the maximum areal extent of a silicic magma chamber, and if a cylindrical shape and a thickness of silicic magma of 1-2 km is assumed, then the maximum volume of silicic magma would be about 30–60 km^3—a modest size. (The assumed thickness is from Smith's [1979, Fig. 12] estimate of the thickness of silicic magma chambers associated with calderas having areas of about 30 km^2.) Consistent with this estimate, Bacon (1985) concluded that a small and comparatively deep magma chamber may be present in the South Sister area, but that a large, shallow magma chamber probably is not. He reasoned that the roof of a large, shallow magma chamber should be decoupled from regional stresses and that overlying vents should have a marked concentric, radial, or diffuse distribution. In contrast, the Holocene rhyodacite vent alignments reflect the regional north-south direction of maximum horizontal compressive stress (Fig. 12). Thus, if a silicic magma chamber of several tens of cubic kilometers is present below the south flank of South Sister, it is too small and (or) too deep to cause significant perturbations in the regional stress field.

In any case, the magma chamber that supplied the Holocene rhyodacite eruptions provided nearly 1 km^3 of magma in two remarkably uniform batches (Tables 3 and 4). This suggests two very different possible sources for the Holocene rhyodacite magmas. These may be viewed as extremes in a wide range of possibilities. (1) The uniform compositions may reflect the tapping of a limited portion of a compositionally zoned magma chamber of much greater volume than the erupted products (e.g., Bacon and others, 1981). Consistent with this possibility, a petrologic model (Hughes, 1983) has suggested that the rhyodacites in the Three Sisters area could have been derived by fractional crystallization from a parental magma of basaltic andesite composition. The small difference in composition between the products of the episodes may reflect the derivation of each batch from a different level in the chamber. The first batch, slightly more fractionated, presumably resided higher in the chamber. (2) Alternatively, the uniform composition of each batch may be due to eruption of an upper-crustal partial melt (e.g., Clark, 1983) that was not greatly fractionated, presumably because of a rapid rate of ascent and eruption. The location of the normative compositions of these lavas between the liquidus field boundaries for 1 kbar—with excess water—and 5 kbar—dry—on a quartz-albite-orthoclase phase diagram (Wyllie, 1977) is not inconsistent with fractionation at upper crustal levels. View 1 suggests that a silicic magma chamber perhaps as large as 30–60 km^3 might be present if the volume estimate discussed above is assumed, whereas view (2) suggests there need not be much of a magma chamber at all. This range of possibilities is obviously large. Future geophysical and petrologic research should be aimed at finding a solution, which is of great importance for both geothermal-energy estimates and for determining the likelihood and potential size of future eruptions.

ACKNOWLEDGMENTS

Supported by the Volcano-Hazards Program of the U.S. Geological Survey. The manuscript benefited greatly from constructive reviews by C. R. Bacon, J. M. Donnelly-Nolan, R. P. Hoblitt, and J. S. Pallister. C. A. Gardner provided the paleomagnetic measurements and helped in their interpretation. I thank L. A. Chitwood of the Deschutes National Forest for valuable discussions and support.

REFERENCES CITED

Bacon, C. R., 1985, Implications of silicic vent patterns for the presence of large crustal magma chambers: Journal of Geophysical Research, v. 90, p. 11243–11252.

Bacon, C. R., MacDonald, Ray, Smith, R. L., and Baedecker, P. A., 1981, Pleistocene high-silica rhyolites of the Coso volcanic field, Inyo County, California: Journal of Geophysical Research, v. 86, p. 10223–10241.

Bailey, R. A., MacDonald, R. A., and Thomas, J. E., 1983, The Inyo-Mono craters: Products of an actively differentiating rhyolite magma chamber, eastern California [abs.]: EOS (American Geophysical Union Transactions), v. 64, p. 336.

Booth, B., Croasdale, R., and Walker, G.P.L., 1978, A quantitative study of 5000 yrs of volcanism on San Miguel, Azores: Royal Society of London Philosophical Transactions, Part A, v. 288, p. 271–319.

Champion, D. E., 1980, Holocene geomagnetic secular variation in the western United States: Implications for the global geomagnetic field: U.S. Geological Survey Open-File Report 80-824, 277 p.

Clark, J. G., 1983, Geology and petrology of South Sister volcano, High Cascade Range, Oregon [Ph.D. thesis]: Eugene, University of Oregon, 235 p.

Crandell, D. R., and Mullineaux, D. R., 1975, Technique and rationale of volcanic-hazards assessments in the Cascade Range, northwestern United States: Environmental Geology, v. 1, no. 1, p. 23–32.

Delaney, P. T., and Pollard, D. D., 1981, Deformation of host rocks and flow of magma during growth of minette dikes and breccia-bearing intrusions near Ship Rock, New Mexico: U.S. Geological Survey Professional Paper 1202, 61 p.

Eichelberger, J. C., Lysne, P. C., Miller, C. D., and Younker, L. W., 1985, Research drilling at Inyo domes, California: 1984 results: EOS (American Geophysical Union Transactions), v. 66, no. 17, p. 186–187.

Fink, J. H., 1984, Spatial variations in the volume, texture, and structure of silicic domes fed by dikes: Geological Society of America Abstracts with Programs, v. 16, no. 6, p. 509.

Fink, J. H., and Pollard, D. D., 1983, Structural evidence for dikes beneath silicic domes, Medicine Lake Highland volcano, California: Geology, v. 11, p. 458–461.

Fisher, R. V., and Schminke, H.-U., 1984, Pyroclastic rocks: Berlin, Springer-Verlag, 472 p.

Heiken, Grant, 1978, Plinian-type eruptions in the Medicine Lake Highland, California, and the nature of the underlying magma: Journal of Volcanology and Geothermal Research, v. 4, p. 375–402.

Hoblitt, R. P., and Kellogg, K. S., 1979, Emplacement temperatures of unsorted and unstratified deposits of volcanic rock debris as determined by paleomagnetic techniques: Geological Society of America Bulletin, v. 90, p. 633–642.

Hoblitt, R. P., Reynolds, R. L., and Larson, E. E., 1985, Suitability of non-welded pyroclastic-flow deposits for studies of magnetic secular variation: A test based on deposits emplaced at Mount St. Helens, Washington, in 1980: Geology, v. 13, p. 242–245.

Hughes, S. S., 1983, Petrochemical evolution of High Cascade volcanic rocks in the Three Sisters region, Oregon [Ph.D. thesis]: Corvallis, Oregon State University, 199 p.

Klein, J., Lerman, J. C., Damon, P. E., and Ralph, E. K., 1982, Calibration of radiocarbon dates: Tables based on the consensus data of the workshop on calibrating the radiocarbon time scale: Radiocarbon, v. 24, no. 2, p. 103–150.

Kuntz, M. A., Spiker, E. C., Rubin, Meyer, Champion, D. E., and Lefebvre, R. H., 1986, Radiocarbon studies of latest Pleistocene-Holocene lava flows of the Snake River Plain, Idaho: Data, lessons, interpretations: Quaternary Research, v. 25, p. 163–176.

Lipman, P. W., 1965, Chemical comparison of glassy and crystalline volcanic rocks: U.S. Geological Survey Bulletin 1201-D, p. D1–D24.

MacLeod, N. S., Sherrod, D. R., and Chitwood, L. A., 1982, Geologic map of Newberry Volcano, Deschutes, Klamath, and Lake counties, Oregon: U.S. Geological Survey Open-File Report 82-847.

Miller, C. D., 1985, Holocene eruptions of the Inyo volcanic chain, California—Implications for possible eruptions in Long Valley caldera: Geology, v. 13, p. 14–17.

Mimura, Koji, 1984, Imbrication, flow directions, and possible source areas of the pumice-flow tuffs near Bend, Oregon, U.S.A.: Journal of Volcanology and Geothermal Research, v. 21, p. 45–60.

Minikami, Takeshi, 1942, On the distribution of volcanic ejecta (Part II): The distribution of Mt. Asama pumice in 1783: Bulletin of the Earthquake Research Institute of Tokyo Imperial University, v. 20, p. 93–106.

Nakamura, K., 1977, Volcanoes as possible indicators of tectonic stress orientation—principle and proposal: Journal of Volcanology and Geothermal Research, v. 2, p. 1–16.

Peterson, N. V., Groh, E. A., Taylor, E. M., and Stensland, D. E., 1976, Geology and mineral resources of Deschutes County, Oregon: Oregon Department of Geology and Mineral Industries Bulletin 89, 66 p.

Pollard, D. D., Delaney, P. T., Duffield, W. A., Endo, E. T., and Okamura, A. T., 1983, Surface deformation in volcanic rift zones: Tectonophysics, v. 94, p. 541–584.

Smith, R. L., 1979, Ash-flow magmatism, in Chapin, C. E., and Elston, W. E., eds., Ash-flow tuffs: Geological Society of America Special Paper 180, p. 5–27.

Sparks, R.S.J., and Wilson, L., 1976, A model for the formation of ignimbrites by gravitational column collapse: Geological Society of London Journal, v. 132, p. 441–451.

Taylor, E. M., 1978, Field geology of the S.W. Broken Top quadrangle, Oregon: Oregon Department of Geology and Mineral Industries Special Paper 2, 50 p.

—— , 1981, Central High Cascade roadside geology, Bend, Sisters, McKenzie Pass, and Santiam Pass, Oregon, in Johnston, D. A., and Donnelly-Nolan, J., eds., Guides to some volcanic terranes in Washington, Idaho, Oregon, and northern California: U.S. Geological Survey Circular 838, p. 55–58.

Thorarinsson, S., 1967, The eruptions of Hekla 1947–1948, I.—The eruptions of Hekla in historical times. A tephrochronological study: Reykjavik, Visindafelag Islendinga, 189 p.

Walker, G.P.L., 1971, Grain-size characteristics of pyroclastic deposits: Journal of Geology, v. 79, p. 696–714.

—— , 1973, Explosive volcanic eruptions—A new classification scheme: Geologische Rundschau, v. 62, p. 431–446.

—— , 1980, The Taupo pumice: Product of the most powerful known (ultraplinian) eruption?: Journal of Volcanology and Geothermal Research, v. 8, p. 69–94.

—— , 1981, Plinian eruptions and their products: Bulletin Volcanologique, v. 44, no. 3, p. 223–240.

Williams, H., 1944, Volcanoes of the Three Sisters region, Oregon Cascades: University of California Publications, Bulletin of the Department of Geological Sciences, v. 27, no. 3, p. 37–84.

Wozniak, K. C., 1982, Geology of the northern part of the southeast Three Sisters Quadrangle, Oregon [M.S. thesis]: Corvallis, Oregon State University, 98 p.

Wozniak, K. C., and Taylor, E. M., 1981, Late Pleistocene summit construction and Holocene flank eruptions of South Sister volcano, Oregon: EOS (American Geophysical Union Transactions), v. 62, no. 6, p. 61.

Wyllie, P. J., 1977, Crustal anatexis: An experimental review: Tectonophysics, v. 43, p. 41–71.

Zoback, M. L., and Zoback, Mark, 1980, State of stress in the conterminous United States: Journal of Geophysical Research, v. 85, no. B11, p. 6113–6156.

Manuscript Accepted by the Society May 5, 1986

Printed in U.S.A.

Geological Society of America
Special Paper 212
1987

Tephra deposits associated with silicic domes and lava flows

Grant Heiken
Kenneth Wohletz
Earth and Space Sciences Division
Los Alamos National Laboratory
Los Alamos, New Mexico 87545

ABSTRACT

Most phases of silicic lava dome growth have some associated explosive activity. Tephra produced during this activity have depositional characteristics, grain sizes, and grain shapes that reflect different mechanisms of dome growth and destruction. It is therefore possible to interpret the explosive history of a dome through study of adjacent tephra deposits even though the dome may no longer be present.

Five stages of dome growth and their associated tephra deposits are considered here. (1) Crater formation before extrusion of a dome, including phreatic, phreatomagmatic (ph-m), and Plinian pumice eruptions, produces a tephra sequence at the base of a dome consisting of deposits rich in accidental lithic clasts from crater walls, overlain by beds of fine-grained tephra and coarse-grained pumice. (2) Magma pulses during dome growth (ph-m, in part) produce tephra consisting of mixtures of juvenile pumice and clasts derived from the partly solidified dome. (3) Ph-m interaction between new magma and a water-saturated dome produces uniform tephra consisting of angular clasts of dome lava. (4) Explosive eruptions that follow collapse of a gravitationally unstable dome produce tephra that consists of angular, partly pumiceous clasts of dome lava which fragment due to expansion of metastable water after release of confining pressure. (5) Posteruptive destruction of the dome by phreatic eruptions results in pyroclasts consisting of fine-grained, hydrothermally altered clasts derived from dome lavas.

Major kinetic processes before explosive dome eruptions are the relatively slow diffusion of magmatic volatiles from magma to fracture planes and foliations within the dome, and the relatively fast diffusion of meteoric water into magma by mechanical mixing. These basic processes control most explosive activity at domes in cases of either expulsion of new magma or collapse of an unstable dome.

INTRODUCTION

Most phases of silicic dome growth have some associated explosive activity. Tephra produced during these explosive events have depositional characteristics, grain-size distributions, and particle shapes that reflect mechanisms of dome growth and destruction. Hopson and Melson (1984) have interpreted Mount St. Helens' pre-1980 history (both constructional and destructional) from a study of older dome remnants and pyroclastic deposits exposed in and around the crater formed in 1980. Explosive activity associated with volcanic dome growth has been analyzed in detail by Newhall and Melson (1983) with regard to history, rate of growth, and petrologic controls of explosions. They point out the "conventional wisdom" that dome growth is a late-stage phase that follows explosive activity. Their analysis of historic eruptions indicates that explosive activity is, however, associated with all stages of dome growth.

Water plays an important role in all explosive activity associated with dome growth and destruction. Dome emplacement is commonly preceded by explosive Plinian and phreatomagmatic (ph-m) eruptions, each with both magmatic and meteoric water as a main volatile component and driving medium. Water is also important in the formation of steam within domes that are destroyed, in part by collapse and subsequent explosions, or by

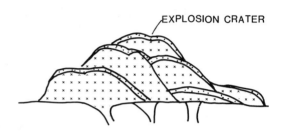
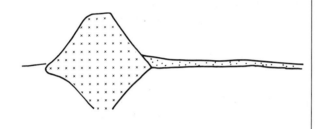
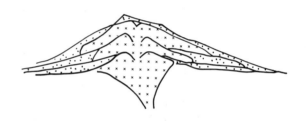

Figure 1. Schematic cross sections illustrating common occurrences of dome-related tephra deposits.

phreatic, crater-forming blasts. Water is also involved in the hydrothermal alteration of dome lavas and in their gravitational collapse because of structural weakening.

On the basis of a review of a number of dome-related tephra deposits, we propose the following eruptive types to classify tephra of similar characteristics: (1) Plinian and phreatomagmatic tephra eruptions preceding dome emplacement; (2) Vulcanian tephra related to periodic and continuing dome growth; (3) Peleean and Merapian tephra associated with dome destruction; and (4) phreatic tephra (from explosively active hydrothermal systems). The tephra of these eruption types occur in a variety of geologic situations (Fig. 1).

We describe tephra by several characteristics, including deposit volume and emplacement mode, tephra-size distributions, constituents, chemical composition, and grain textures as determined by scanning electron microscopy (SEM). The SEM studies are emphasized because they present the most distinguishing data for tephra characterization.

PLINIAN PUMICE AND PHREATOMAGMATIC DEPOSITS THAT PRECEDE DOME EMPLACEMENT

Most silicic domes are extruded at the end of an eruption cycle that begins with highly energetic, gas-rich eruptions that produce Plinian pumice deposits and associated pyroclastic flows. If there is ample ground or surface water close to the vent, phreatomagmatic activity can produce fine-grained silicic tephra that are present both as a widespread deposit and proximal tuff ring through which the dome is later extruded (Fig. 2). Several examples are used here, including Plinian pumice deposits in the Medicine Lake Highland, Chaos Crags, and Panum Crater, California. These deposits have small volumes (<0.5 km^3), are associated with fissure eruptions of silicic magma, and are overlain by silicic domes.

Little Glass Mountain and Glass Mountain

Models for dome formation and growth were developed by

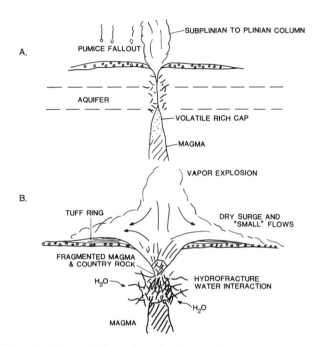

Figure 2. Schematic illustration of tephra production preceding dome formation with a Plinian stage (A) followed by phreatomagmatic explosions (B).

TABLE 1. SUMMARY OF HOLOCENE SILICIC ERUPTIONS, CHAOS CRAGS AND MEDICINE LAKE HIGHLAND, CALIFORNIA*

	Chaos Crags	Medicine Lake Highland
Total volume	1.4 km³	1.2 km³
Tephra volume	10% of total	10% of total
Main mode of tephra deposition	pyroclastic flow	Plinian ash flow
Phenocryst content of pumice	30-40%	0-trace
Pumice characteristics: Vesicularity	14-30%	50-60%
Predominant vesicle shape	ovoid; little coalescence	highly elongate; coalesced vesicles
Glass composition	rhyolite	rhyolite
Mixed magma	>90%	10%

*From Heiken and Eichelberger (1980).

Fink (1983) at the Little Glass Mountain and Glass Mountain, California, rhyolite flows. Emplacement of each dome complex was preceded by a Plinian eruption. Pumice-fall deposits from these eruptions range in thickness from 4 to 6 m near the vent, to a few cm at a distance of 25 km from the source (Table 1). There are only small pumice rings developed around vents. Most of the pumice beds consist of multiple, 20- to 50-cm-thick, reversely graded beds of pale-gray pumice lapilli and coarse ash (Heiken, 1978). Pumice beds drape underlying terrain and exhibit no flowage structures. All tephra units consist of very poorly sorted lapilli and are within the grain-size field of tephra-fall deposits as described by Walker (1971).

Pumice pyroclasts (Fig. 3A) are equant to slightly elongate and have sharp, angular surfaces. Most have 40% to 50% vesicles; vesicles are 10 to 500 µm long and are highly elongate. Pumices from both vents (Glass Mountain and Little Glass Mountain) are remarkably homogeneous throughout the eruption sequence. Blocky grain surfaces are both parallel and normal to the long axes of vesicles. Most pumices contain some cavities that consist of partly collapsed vesicles surrounded by elongate vesicles. All pumices consist of aphyric glass; only samples from Little Glass Mountain contain traces of orthopyroxene phenocrysts.

Pumice pyroclasts in these deposits are highly vesicular and lack phenocrysts, indicating rapid rise of magma along dikes. All pumiceous pyroclasts have parallel vesicles with high aspect ratios. Vesicles appear to have formed prior to eruption and were stretched during flow, parallel to dike walls. Angular fractures are oriented parallel and normal to elongate vesicles and form a distinctive type of pyroclast; these may have formed by disruption of vesiculating magma by passage of a rarefaction wave downvent. Fragmented, vesicular clasts were then accelerated out of the vent (an idea that was first proposed by Rittman, 1936). After the volatile-rich top of the magma body was erupted as tephra, the bulk of dome lavas with low (<20%) vesicularities were erupted (Fink, 1983). A total volume of 1.2 km³ was erupted during this Holocene rhyolitic event; 10% of that volume (dense-rock equivalent) consists of pumice deposits erupted before the domes (Heiken, 1978).

Chaos Crags, California

The Chaos Crags are a line of Holocene age dacite domes located immediately north of Lassen Peak, California. Prior to eruption of these domes, explosive activity produced ash falls and pyroclastic flows of small volume (Table 1). The pyroclastic-flow deposits consist of massive, nonwelded coarse ash that, in some places, contain large pumice lapilli and blocks. Beds with more than 10% blocks and bombs are reversely graded. Ash-fall beds consist of massive and normally graded medium to fine ash and lapilli-bearing coarse ash (Heiken and Eichelberger, 1980).

In contrast to the aphyric, highly vesicular pumice pyroclasts of Glass Mountain and Little Glass Mountain, Chaos Crags pumices are porphyritic and poorly vesicular. Vesicularities range from 6% to 46%, but the average is about 30%. Vesicle shapes range from spherical to highly elongate, but most are ovoids, 10–200 µm long. Pumices contain between 30% and 45% anhedral to subhedral phenocrysts of plagioclase, hornblende, biotite, quartz, and magnetite. Many vesicles in these pyroclasts radiate from phenocryst surfaces.

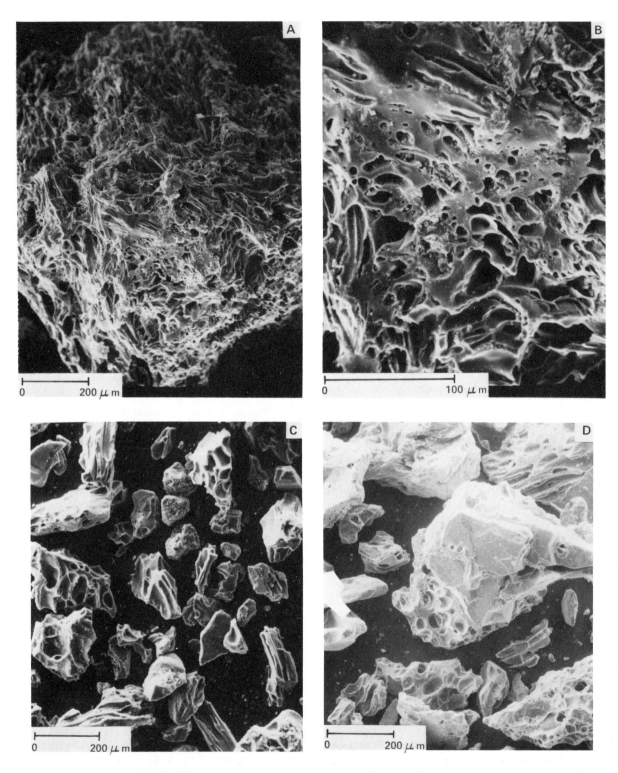

Figure 3. Scanning electron micrographs of pumice pyroclasts from Glass Mountain, Medicine Lake Highland, and Chaos Crags, California. *Glass Mountain.* A: Angular, subequant rhyolitic pumice. It is all glass and contains no phenocrysts. This is an "end-on" view of the pyroclast; short axes of highly elongate vesicles are shown. Larger cavities are composite and were formed by coalescence of adjacent vesicles. B: Detail of A showing the variation in vesicle widths. Most vesicles are thin walled, and many exhibit some degree of coalescence. *Chaos Crags.* C: This tephra consists of a mixture of pumice and crystal pyroclasts. Pumices are ash size and angular, and contain mostly thick-walled, ovoid vesicles. D: Another tephra pumice and phenocrysts are coated with pumiceous glass.

Tephra deposits

PANUM CRATER

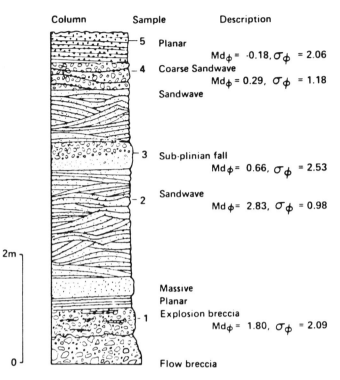

Figure 4. A stratigraphic section of the basal part of the Panum Crater tuff ring showing the sequence of mainly pyroclastic surge deposits with one subplinian pumice bed. The upper portion of the section (not illustrated) consists of repetitions of planar bedding shown at sample interval 5. Note the tephra size parameters for each studied sample.

TABLE 2. CHEMICAL ANALYSES OF PANUM CRATER PYROCLAST SURFACES IN WT%*

	1	2	3	4	5
SiO_2	67.72	72.45	69.90	68.45	70.83
TiO_2	2.14	1.64	2.02	2.22	1.79
Al_2O_3	13.49	12.62	12.92	12.86	13.02
FeO	3.692	1.97	2.64	3.13	1.29
MnO	1.86	1.26	1.62	1.98	1.40
MgO	1.59	1.30	1.54	1.51	1.48
CaO	2.40	1.85	2.23	2.34	2.01
Na_2O	1.37	1.39	1.67	1.50	1.52
K_2O	5.80	5.55	5.94	6.01	5.77
Glass	85.9	74.0	83.0	87.0	83.0
Crystal	4.7	24.0	12.5	8.0	12.0
Lithic	9.4	2.0	4.5	4.8	4.0

*Normalized standardless energy dispersive spectral analyses of grain surfaces.

Pumice from the Chaos Crags eruption have low vesicularities; lower volatile contents may have been related to crystallization of a hydrous phase (Fig. 3b). Vesiculation appears to have occurred near surface where there was little chance of vesicle elongation by flow. As in the Medicine Lake Highland, tephra from Chaos Crags make up about 10% of the total volume erupted (dense-rock equivalent; DRE); 1.1 km³ is lava erupted as a stubby flow and a three-lobed dome.

Panum Crater, California

Panum Crater is a Holocene-age rhyolite dome surrounded by a tuff ring 600–700 m in diameter. It erupted at the northern end of the Mono Crater chain in east-central California (Putnam, 1938; Kistler, 1966; Sieh, 1983). Its opening explosive eruptions have been dated by Wood (1977) at 1190 ±80 yr B.P.

To illustrate the sequence of pyroclastic material constituting the tuff ring we include here a stratigraphic section measured in a pumice quarry on the ring's southeast flank (Fig. 4). The initial explosive eruptions produced an explosion breccia, which is overlain by dominantly dune-bedded pyroclastic surge beds with an intercalated sub-Plinian pumice- and ash-fall layer. The last explosive activity, prior to dome extrusion, deposited planar-bedded surge deposits that formed upper portions of the ring. The volume of these tephra is estimated to be between 0.005 and 0.05 km³.

Pyroclast size data, shown in Figure 4, display a gradual coarsening of tephra upward in the section from a median diameter of about 150 µm to 1 mm. The dune-bedded materials are well sorted, whereas the fall and planar-bedded surge deposits are moderately to poorly sorted.

Phenocryst-poor rhyolite magma was erupted at Panum Crater. Grain surface alteration is slight on all samples except those of the lower strata, where coatings and vesicle fillings of altered glass and claylike material are visible. Chemical changes on tephra surfaces, caused by rapid posteruptive surface alteration, is typical of phreatomagmatic eruptions (Table 2) (Heiken and Wohletz, 1985). Accidental lithic pyroclasts constitute nearly 10% of early explosion materials and about 5% of later tephra. Variation of phenocryst abundance throughout the eruption sequence is also evident (Table 2). This variation and the degree of glass alteration (i.e., deviation from expected rhyolite compositions), explain differences in chemical compositions throughout the eruption sequence.

Surface textures (Fig. 5) of vitric pyroclasts are dominated by blocky and equant shapes and minor to moderate development of vesicles (Table 3). Where present, vesicles are cut by planar to curviplanar fracture surfaces; fine dust particles adhere preferentially to vesicle surfaces. Many vesicles, especially those of later erupted tephra, are elongate and range from parallel tubelike hollows to flattened ellipsoids that define a foliation that is also apparent in later extruded lavas. Overall, vesicle abundance appears to decrease upward in the section. Abrasion features on grain surfaces are poorly to moderately developed and consist of some rounding of initially sharp edges, chips, grooves, and conchoidal-like fracture planes.

Figure 5. Stratigraphic column (top) and scanning electron micrographs (SEM) (bottom) of Panum Crater tephra. SEM show typical features of silicic, dominantly phreatomagmatic ash. A: Poorly vesicular slablike grain from the explosion breccia, sample 1. Vesicle is filled with fine ash and some alteration products, and the grain surface is coated with hydrated glass. B: Blocky, vesicular grain from sandwave sequence, sample 2. This grain has several planar fracture surfaces and edges are partly rounded by transport abrasion. C: Pumiceous equant grain from Subplinian bed, sample 3. Although vesicular and displaying several surfaces bounded by curved vesicle walls, this grain is also somewhat blocky, which probably indicates fragmentation in part by quench stresses during phreatoplinian eruption. This sample best matches Wood's (1977) tephra #2 that is widely distributed over the southeastern Sierra Nevada. D: Angular, blocky ash grain from the upper portion of the measured section, sample 5; the poor vesicularity, rounded corners, and surface pits, chips, and scratches typify ash formed by late-stage phreatomagmatic fragmentation of the viscous, volatile-poor magma erupted just prior to passive lava dome extrusion.

Tephra produced by the opening eruptions are dominantly phreatomagmatic and subordinantly magmatic phreato-Plinian, because vesiculating rhyolite magma was extruded through water-saturated alluvial deposits present around Mono Lake.

VULCANIAN ERUPTIONS ASSOCIATED WITH DOME GROWTH

Vulcanian activity encompasses a wide range of eruption phenomena; this range includes tephra emission during formation of silicic domes and plugs at polygenetic volcanoes (Fisher and Schmincke, 1984). These volcanoes may be dome complexes or composite cones that display periodic and, in some cases, cyclic tephra emissions. This activity has its name derived from the Aeolian island Vulcano, in the Tyrrhenian Sea, where Mercalli and Silvestri (1891) witnessed and described the 1888–1890 eruption of the Fossa cone. In that eruption, intermittent, staccato, and cannon-like bursts of ash and accretionary lapilli

TABLE 3. TEXTURAL FEATURES OF PANUM CRATER TEPHRA PARTICLES

Sample and Bed Form*	Grain Shape	Edge Modification	Alteration	Fine Fraction (wt.%)	Distinguishing Features
1 M	vesicular, blocky	subangular smooth rounded vesicle edges	moderate with vesicle fillings	blocks and plates, some adhering dust (9%)	vesicular with lithic material
2 SW	blocky	subangular, grooves, dish shape fractures	slight, mostly clean surfaces	very thin plates, abundant micrometer-size dust (11%)	impact fractures
3 F	blocky, vesicular	angular, none	slight, clean surfaces	blocks and plates, no fine dust (5%)	no abrasion
4 SW	blocky, vesicular	subrounded, stepped and dish shape fractures	slight, vesicle fillings	blocks, adhering dust (1%)	highly abraded edges
5 P	blocky	angular, chipped edges	none	blocks, no adhering dust (2%)	clean surfaces, few vesicles

*Bed forms are: SW - sandwave, M - massive, P - planar, F - fall.

that produced bedded ash falls, surges, and lahars on the slopes of the Fossa cone were followed by quiet periods of several minutes to several days. Early eruptions ejected dominantly accidental lithic ash derived from old vent lavas. Later explosions produced juvenile blocks, breadcrust bombs, and ash.

Much Vulcanian activity can be attributed to ground water in contact with new magma below the cone, whereas later explosions are caused by vesiculating magma (Fig. 6). The three examples described below include tephra associated with renewed activity at Vulcano in 1888 and continuing dome growth at Mount St. Helens, Washington, and Santiaguito, Guatemala in recent years.

Vulcano, Italy; 1888–1890 Eruption

Recent studies by Sheridan and others (1981) and Frazzetta and others (1983) carefully documented pyroclastic materials deposited during the Vulcanian eruptions of 1888–1890. By comparing these deposits with those of earlier eruptions, these workers illustrated the clear recurrence of Vulcanian activity. Figure 7 is a stratigraphic section of the four most recent eruption cycles and shows a progression for each cycle, from explosion breccia (partly destroying the previous plug) to ash fall to wet and dry surge eruptions; these are followed by pumice falls and lava-flow extrusion. Much of the tephra deposited on volcano slopes was remobilized as lahars and moved off cone slopes.

Typical tephra deposits consist of ash and lapilli falls, sets of thinly bedded coarse and fine ash, dry surge dunes and planar beds, and wet-surge beds that are transitional to lahars. The total volume of tephra from the 1888–1890 eruption cycle is estimated to be in the range of 0.01 km^3. Most samples are characterized by two size populations with dominant modes at 0.5 to 1.5 ϕ and

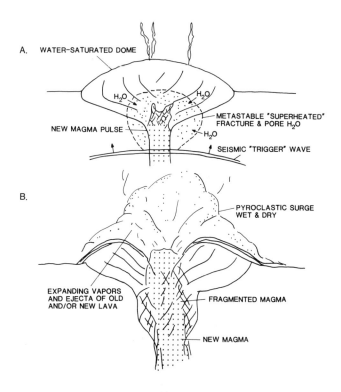

Figure 6. Schematic illustration of preeruptive (A) and eruptive (B) stages for an example of Vulcanian activity.

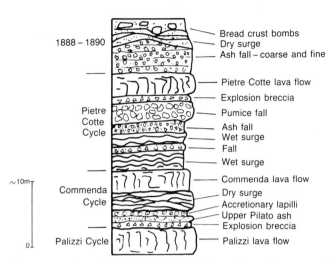

Figure 7. Stratigraphic columnar section adapted from Frazzetta and others (1983) of the past four eruptive cycles at Vulcano. The most recent cycle, 1888–1890, typifies Vulcanian activity associated with silicic lava extrusion; samples from this deposit are described.

2.5 ϕ to 3.5 ϕ; however, ash and lapilli falls commonly show an additional mode in the range of -2.0 ϕ to -1.0 ϕ. Wet-surge deposits are typically very fine grained and have a mode finer than 4.5 ϕ. The significance of tephra size modes has recently been discussed by Frazzetta and others (1983, 1984), who ascribed them to traction, saltation, and suspension modes of transport.

On the basis of tephra petrography, Frazzetta and others (1983) suggested that mixing of evolved aphyric rhyolite and tephrite magmas occurred at Vulcano. A variety of phenocrysts are noted in the tephra. Phenocrysts of clinopyroxene, calcic plagioclase, and olivine are entirely free of glass coatings and have cleavage faces modified by stepped fractures and some shallow, dish-shaped concavities (Fig. 8A). Glass particles are poorly vesicular to nonvesicular (Fig. 8B) and have blocky shapes with sharp, acute edges and broad curviplanar surfaces. Abrasion has rounded some grain surfaces; these surfaces were also pitted and coated during postdepositional solution and precipitation. Rare pyroclasts with fluidal and ropy surfaces formed from fragmented portions of the mafic-melt fraction.

Many of the glassy tephra are accidental lithic clasts derived from older dome lavas. This is characteristic of Vulcanian tephras, which generally display marked surface hydrothermal alteration that produces a coated or "muddy" surface texture.

Mount St. Helens; Continuing Dome Growth

Considerable scientific attention on the continuing activity at Mount St. Helens since 1980 has greatly increased our awareness of the complexities of dome growth (see Swanson and others, this volume; Hopson and Melson, 1984). This activity has taken place over 6 years as intermittent eruptions of dacite lava flow lobes of small volumes (1–4 × 10^6 m^3); tephra emissions associated with some of these magma pulses produced plumes several thousand metres high and small pyroclastic flows and lahars (Swanson and others, 1983; Waitt and others, 1983). Tephra volumes are much less than those of the associated lavas. These authors have noted some correlation of the amount of snowpack present and the magnitudes of small explosions.

The tephra samples discussed here were provided by Don Swanson, who collected them in April 1983 from snow on the crater floor about 400 m from the vent. Tephra characteristics discussed here are solely those obtained from SEM observation. Coarser materials were carbon coated for low resolution (50 to 1000 X) imaging and semiquantitative energy dispersive spectral (EDS) analysis. Fine fractions were gold coated for high resolution (3000 to 30,000 X) imaging.

The samples are fine- to medium-ash size; the largest clasts are in the range of 500 to 900 μm and there is abundant material in the size range of 50 to 125 μm; the finest material is 0.2- to 1.0-μm-size dust attached to 1 to 15 μm particles.

Textural features (Fig. 9) of the ash are presented in Figure 10 as percentage abundance of a textural feature in each sample (from grain counts of scanning electron micrographs). These data are plotted as a function of stratigraphic position and reveal tephra properties related to eruptive mechanism as discussed by Wohletz and Krinsley (1982). Several features are readily apparent and can help in understanding tephra production related to dome growth (Fig. 10): (1) vesicularity is generally low (<10%) for coarse particles but higher (10–70%) for fine ones; (2) blocky, equant grain shapes have an inverse relationship to vesicularity (pyroclast textures are dominantly blocky); (3) fused and fluidal grain surfaces are more apparent in the fine fraction than in the coarse fraction, but the opposite relationship is noted for grain-surface alteration. Fused surfaces indicate something about the viscosity, and hence, the temperature of the fragmented lava; (4) fine adhering dust is noted on about 40% of all particles and is most prevalent on fine-grained particles. The finest ash is most abundant in eruptions of the most fluidal and hottest magmas, and (5) the abrasion of coarse particles is low (15–30% of the particles) and most likely represents reworking of accessory fragments derived from pulverization of dome lavas and older tephra. Accessory grains generally appear to be the most hydrothermally altered and pose a problem for SEM viewing because their surfaces collect a substantial charge under the electron beam.

Semiquantitative measurement of major-element abundances on grain surfaces is illustrated in Figure 11. The variation of elemental abundances among samples might seem surprising; however, it simply shows that the amount of grain surface alteration is strongly controlled by hydrothermal processes within the dome before and during each ash-eruption pulse. The analyses are compared with average values for the 1980 to 1982 dome

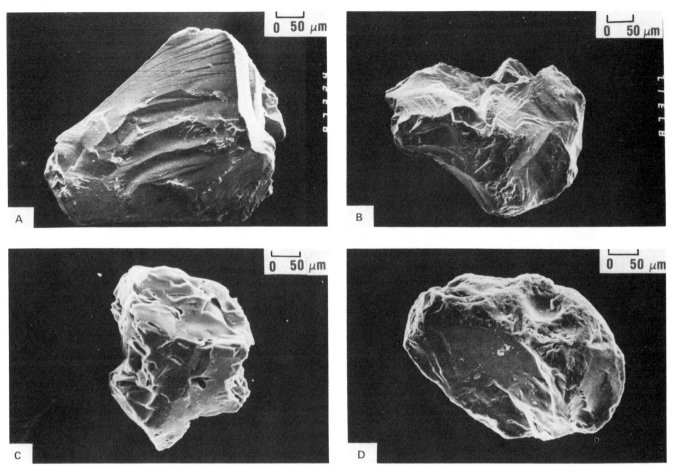

Figure 8. A, B: Scanning electron micrographs (SEM) of crystal pyroclasts from Vulcano. Note cleavage faces modified by stepped fractures. C, D: SEM of glass pyroclasts from Vulcano. These poorly to nonvesicular grains show some abrasion rounding, chipping, and pitting.

lavas and pumice of the AD 1500, 1800, and 1980 eruptions reported by Cashman and Taggart (1983) and Melson (1983). In all cases, the average composition found for all tephra surfaces is very close to that of the dome lavas. It is interesting to note that these analyses vary with samples in a similar fashion as do textural features discussed above. If the limitations of EDS analysis of particle surfaces are considered, it is difficult to know the exact meaning of the values obtained; however, we have noted similar variations of tephra surface chemistry in a number of tephra sequences studied from other volcanoes (Heiken and Wohletz, 1985). A conservative interpretation is that varying eruption processes within the dome are reflected in the degree of hydrothermal alteration of tephra. Analytical values (especially SiO_2) are both the most similar and the most dissimilar to those of pumice from past explosive eruptions for samples that show the greatest phreatomagmatic textural components of blockiness and surface alteration. Vesicularity is greatest for samples most similar to those of dome lavas.

Our interpretation of this complex data set is mostly independent of any field data. We believe that both magmatic and phreatomagmatic mechanisms play a part in eruption of tephra from the dome at Mount St. Helens, and that the relative dominance of these mechanisms alternate sequentially with time. Overall tephra vesicularities are relatively low when compared with tephra of magmatic eruptions. Therefore, phreatomagmatic eruptions most likely produced most of the tephra and may be controlled by the rate of ground-water flow derived from snowmelt and rainfall and the rise rate of new magma. In this situation, phreatomagmatic activity encompasses direct contact of fluid magma with ground water and the development of an unstable hydrothermal system around congealing subsurface magma. Because vesicularity, adhering fine dust, and fused surfaces are most prevalent in the fine-fraction tephra, these particles likely represent the magmatic component of the samples, whereas coarse-fraction materials reflect more of the phreatic accessory and phreatomagmatic constituents.

Santiaguito, Guatemala

Santiaguito is a dome complex in southwest Guatemala that

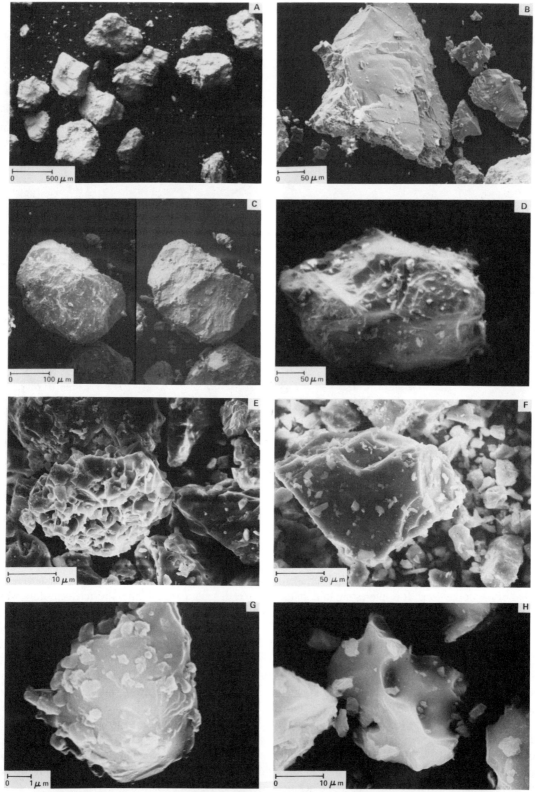

Figure 9. Scanning electron micrographs of tephra erupted from the Mount St. Helens dome. A–D: Coarse particles showing blocky texture, strong surface alteration (C shows both secondary electron and backscattered electron images), and scalloped, rounded edges. E–H: Fine-fraction particles show some vesicularity and abundant adhering dust. Surfaces are mostly fresh in appearance, and grain angularity is high.

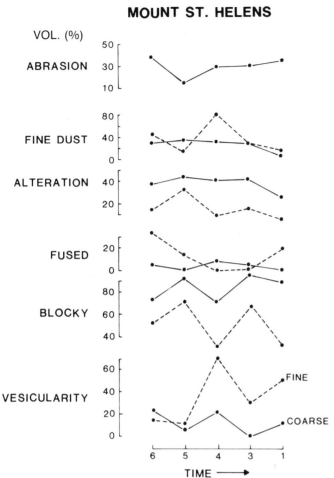

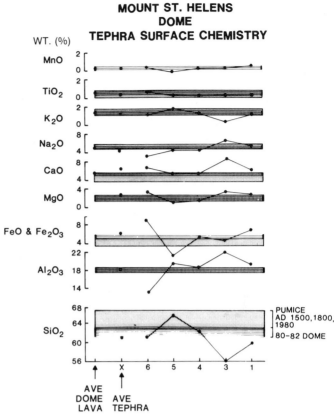

Figure 10. Textural features of tephra from the Mount St. Helens dome. Abundances of the features are obtained from grain counts for coarse and fine (<50 μm) ash fractions and are plotted vs. stratigraphy.

Figure 11. Major element chemistry of tephra surfaces from Mount St. Helens showing variations of surface alteration with stratigraphy. Average values are shown for all tephra, for the 1980–1982 dome lavas, and pumice from the AD 1500, 1800, and 1980 explosive eruptions (Cashman and Taggart, 1983; Melson, 1983). Note that the tephra analyses are normalized to total 100 wt.% and are standardless, semiquantitative EDS results.

consists of dacite domes and flows; it has been erupting since 1922 (Rose, 1973; Rose, this volume). Sporadic, small eruptions from craters developed over the Caliente and El Brujo vents produce tephra falls and nuée ardentes. The activity can be classified as Vulcanian. Tephra from these eruptions consist of mostly hyalocrystalline, equant, blocky dacite fragments (Fig. 12). Grain surfaces are hackly to conchoidal and have low vesicularities. There are 20% to 30% crystal pyroclasts, including plagioclase and lesser amounts of quartz. Rose (this volume) reports that near the vent there are bread-crust bombs mixed with lithic ash.

PELÉEAN AND MERAPIAN ERUPTIONS: TEPHRA FORMED BY DOME DESTRUCTION

Explosive disruption or destruction of silicic domes produces tephra composed primarily of poorly vesicular lithic pyroclasts derived from partly to completely solidified lavas. The main example used here is from the 1902 eruption of Mont Pelée, Martinique. LaCroix (1904) described the eruption as having been laterally directed, whereas Fisher and others (1980) and Fisher and Heiken (1982) interpreted the eruption column as having been vertical, followed by column collapse. In either case, the dome was destroyed and produced a small volume of highly destructive block and ash flows and rapidly moving, high-energy ash clouds. At Merapi, Java, in 1930, part of the dome collapsed and produced pyroclastic flows in which about 10% of the clasts were juvenile (Neumann van Padang, 1931). What all of these eruptions have in common are tephra that consist of mostly lithic pyroclasts formed by fragmentation of dome lavas (Fig. 13).

Mont Pelée, Martinique, 1902 Eruption

The eruptions of May 8 and 20, 1902 were of the most destructive explosive phases of the Mont Pelée eruption. Recent analyses of the pyroclastic deposits indicate that eruptions were

Figure 12. Tephra fall from Santiaguito, Guatemala, 1970 eruption. These are mostly angular lithic clasts that consist of poorly vesicular, hyalocrystalline dacite.

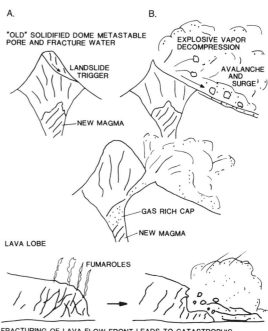

Figure 13. Schematic illustration of Peléean and Merapian dome destruction, showing initial landslide (A) followed by avalanching and dome explosion (B) caused by explosive release of new, gas-rich magma (center), and explosion of lava flow front (bottom).

vertically directed and that pyroclastic flows were generated by collapse of the eruption column (Fisher and others, 1980; Fisher and Heiken, 1982). There were two parts to these flows; a lithic-rich fraction derived from the dome collapse, which flowed out of a notch in the crater, and a still-fluid portion that may have erupted around the solidified neck and dome. Gravity segregation of coarse lithic fragments concentrated them as block and ash flows in the valley of Rivière Claire and as turbulent, high-energy ash clouds that spread over the landscape.

Tephra from these eruptions are mostly blocky, tabular hyalocrystalline andesite clasts (Fig. 14). Most contain feldspar and pyroxene microlites and some ilmenite and feldspar phenocrysts in a glassy groundmass (80% by volume). Vesicles are rare; vesicularities range from zero to a trace of 1–10 µm-long vesicles, to 10% in clasts having diktytaxitic textures. Coarser blocks and lapilli are similar in appearance and composition. Most of the tephra could have been produced by the breaking up of dome lavas. Glass shards and pumices are present (1% to 50%) and are most abundant in the <63 µm size fraction; they range from simple angular shards to pumices with 60–80% vesicles; most pumices have 4- to 15-µm-long, ovoid, thin-walled vesicles (Fisher and Heiken, 1982). These represent the molten fraction of magma that reached the surface.

Pressure leading to explosive fragmentation of the dome at Pelée could have been caused by magmatic gas, vaporization of ground water, or both. Evidence presented by Fisher and Heiken (1982) is ambiguous as to the phreatomagmatic component of the eruption. Phreatic activity may have triggered the activity, as heavy rains occurred the night before the activity of May 8. Westercamp and Traineu (1983), in their analysis of the past 5000 years of explosive activity at Mont Pelée, noted that a major type of eruption (one of four types described) is dome growth and subsequent explosive fragmentation and collapse that feeds glowing avalanches and block and ash flows. This type of activity has occurred during about 60% of eruptions at Mont Pelée.

Differences between Peléean and Merapian activity appear to be little more than timing. Peléean activity occurs during or immediately after dome emplacement, whereas Merapian collapse occurs at some unspecified time after dome emplacement, but while there is still some hot dome lava in the conduit. Tephra deposits and pyroclast characteristics are very similar and may differ only in the amount of juvenile pumice present.

Variations on this tephra type are those associated with explosive disintegration of the foot of a silicic lava flow. Rose and others (1976) have described production of pyroclastic flows following collapse of the nose of a blocky dacite flow (El Brujo flow) at Santiaguito, Guatemala, in 1973. Total volume of erupted material was less than 200,000 m^3. Pyroclasts in this deposit are broken crystals, lithic clasts, and vesicular and nonvesicular glass.

PHREATIC TEPHRA

Phreatic eruptions produce accidental lithic tephra during

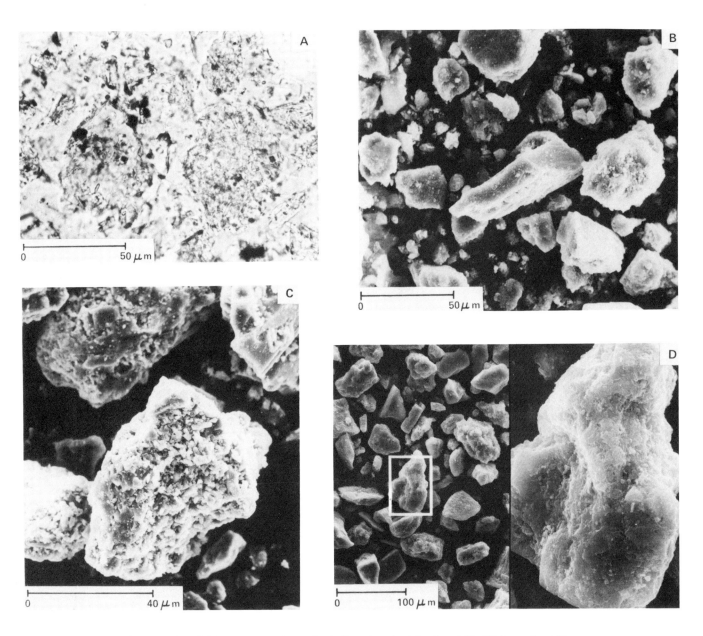

Figure 14. Samples from surge and pyroclastic flow deposits. May 8 and 20, 1902 eruptions, Mont Pelée, Martinique. All samples used here are from the fine-grained matrix of these deposits. A: Photomicrograph (transmitted light) of matrix from block-and-ash flows, unit 1 (Fisher and Heiken, 1982). Over 70% of this sample consists of slightly altered to fresh, angular clasts of hyalocrystalline andesite. B: Scanning electron micrograph (SEM) of matrix from the base of the earliest block and ash flow. Seventy percent of the pyroclasts consist of andesitic lava. Surfaces are irregular and slightly altered. There are also some plagioclase clasts (center). C: SEM of lithic pyroclasts, uppermost block and ash flows. Clast surfaces are very irregular, but angular, with poorly developed diktytaxitic textures. There are traces of pumice in this sample, which is from the uppermost part of the eruption sequence. Smaller pyroclasts cling to grain surfaces, a characteristic of tephra from pyroclastic flows. D: SEM of pyroclasts from the uppermost surge deposits. The deposit consists of 60% hyalocrystalline andesite clasts and 28% phenocrysts (mostly plagioclase). Lithic clasts range from angular to subrounded (inset). There are 1% shards and pumice in this deposit.

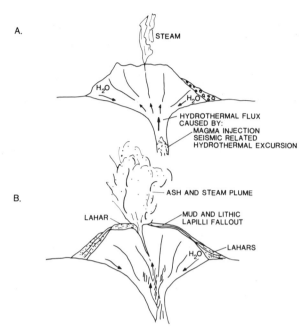

Figure 15. Schematic illustration of phreatic activity where dome is not destroyed. Passive fumarolic stage (A) is followed by energetic steam releases with entrainment of lithic tephra (B).

energetic fumarolic activity and steam explosions associated with hydrothermal processes in domes. This activity may result in slight vent widening and mantling ash and lapilli layers and lahars (Fig. 15), or complete destruction of a dome edifice by steam explosions caused by a new magma pulse beneath the dome.

La Soufrière de Guadeloupe, French West Indies

La Soufrière de Guadeloupe is a steep-sided andesite dome that has had a historic record of explosive phreatic activity. The last such activity occurred during late summer and fall of 1976 and caused evacuation of thousands of residents of the island of Basse-Terre, Guadeloupe. Explosive steam eruptions from open summit fissures and along the southeast flanks of the dome entrained fine-grained tephra that were deposited downwind south and west of the volcano as ash fall and very dilute density currents (Heiken and others, 1980). These tephra formed a gray, thin, sticky deposit that contained accretionary lapilli. During larger eruptions, thin streams of light-gray, steaming mud flowed down dome flanks as lahars (Wohletz and Crowe, 1978).

Tephra from the best-observed phreatic eruption (October 2, 1976), collected about 2 km from the vent, is poorly sorted (σ_ϕ = 2.03) and had a mean grain size of 38 μm (mean grain size [M_z] = 4.7 ϕ). The finer (<38 μm) fraction is composed of lithic and

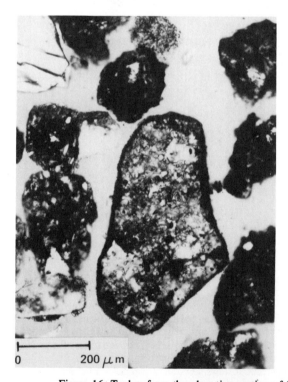 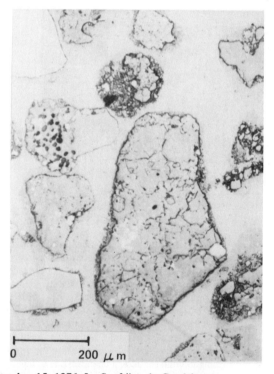

Figure 16. Tephra from the phreatic eruption of September 15, 1976, La Soufrière de Guadeloupe (>38 μm fraction). Left: Transmitted light photomicrograph, partly crossed polarizers. Right: Reflected light photomicrograph. The sample consists of mostly subangular altered lithic pyroclasts; finely crystalline, equigranular lavas derived from dome lavas. Alteration products within lithic clasts include pyrite (bright phases in reflected light), hematite, and clay. Many of the clasts are coated with thin layers of clay.

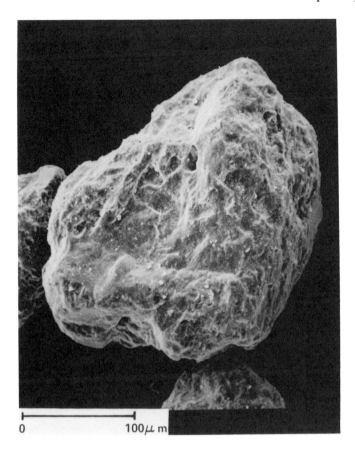 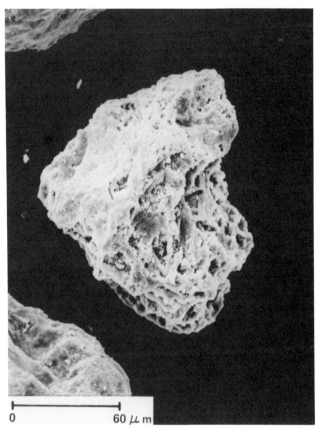

Figure 17. Scanning electron micrographs of lithic clasts from the La Soufrière de Guadeloupe phreatic eruption of August 26, 1976. There are mostly hydrothermally altered, subequant clasts of dacitic lava. All clasts contain pyrite and are partly altered to clay.

mineral fragments, sulfates, sulfides, authigenic silica, and oxides. The coarser fraction was used for modal analysis.

Lithic pyroclasts consist of equant to subequant, subangular to subrounded dacite or andesite fragments that exhibit a wide range of alteration. The most common lithic clasts consist of stubby plagioclase or orthopyroxene phenocrysts in a groundmass of microcrystalline feldspar and pyroxene phases or colorless glass (Figs. 16 and 17). The degree of alteration of these clasts ranges from those that are nearly fresh and have only a trace of hematite stain, through those laced with hematite or pyrite, to those that consist of mixtures of authigenic silica, pyrite, and anhydrite.

Mineral pyroclasts include plagioclase grains that have compositions of An_{50} to An_{88}, and are present as broken, subhedral crystals. As was the case for lithic clasts, mineral phases range from "fresh" (slightly etched grain surfaces) to ragged skeletons. Pyroxene pyroclasts are pale-yellow to pale-pink orthopyroxenes ($Wo_3En_{63}Fs_{34}$) (Fig. 18). Some of the fresher grains contain 20- to 200-μm-long, brown-glass inclusions. Traces of clinopyroxene are present. Amorphous masses of hematite and 30- to 50-μm pyrite grains are present in small amounts in many of the samples and as trace amounts in all samples. There are also traces of anhydrite and tridymite.

Most glass pyroclasts are angular, equant, slightly vesicular to nonvesicular fragments that range in composition from andesite to rhyolite (Table 4). Most appear to have been derived from the glassy groundmass of hyalocrystalline dome lavas. Rare colorless glass pyroclasts have 10% to 20% ovoid vesicles (long axes of 10 to 15 μm); nearly all are partly altered and do not represent fresh juvenile pyroclasts. Also present are rare orange-brown basaltic glass fragments that have palagonitic rims.

Tephra from the 1976 phreatic eruptions of La Soufrière de Guadeloupe have been derived from hydrothermally altered and acid-leached dome rocks and older lavas. Lithic pyroclasts and lavas are similar in texture and composition. An abundance of slightly altered to nearly fresh plagioclase grains in the tephra may be related to preferential alteration of the fine-grained or glassy groundmass of the lavas. None of the colorless glass pyroclasts were fresh enough to classify as juvenile tephra. The activity at La Soufrière appears to have been that of a vapor-dominated geothermal system driven by a localized heat source in the form of a magma body located at relatively shallow depths. New magma at

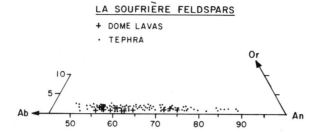

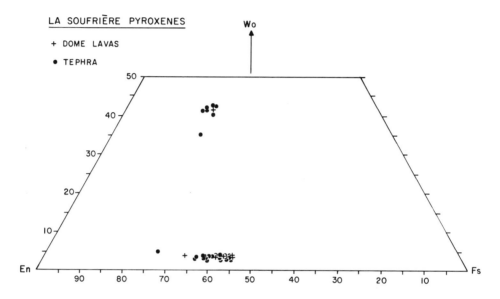

Figure 18. Mineral compositions, Soufrière de Guadeloupe. Texturally, the clasts within the phreatic tephra appear to have been derived from the lava dome. Overlapping mineral compositions (pyroxenes and feldspars) also demonstrate this similarity.

depth was suggested by greatly increased seismic activity and shallow epicenters (M. Feuillard, 1977, personal commun.).

Phreatic (violent steam) eruptions without the eruption of magma are common. This type of activity has preceded all magmatic eruptions at Mont Pelée, and often occurs alone (Westercamp and Traineu, 1983). Vapor-dominated hydrothermal systems are often established within silicic domes, where water recharge is at the summit (Healy, 1976). La Soufrière is the highest mountain on Guadeloupe (elevation 1466 m); it consists of fractured, highly permeable lavas, and is the center of high rainfall (Lasserre, 1961). This geologic setting fits models of ground-water movement described by Healy (1976), in which high-temperature fumaroles are present at the summit (1466 m) and warm springs issue from the base (950 to 1200 m elevation) of a volcanic dome. Increased circulation within this system by movement of magma at depth and seismic activity can cause explosive eruptions without magma reaching the surface. Explosive activity was also stimulated after heavy rainfall. Heiken and others (1980) suggested that vapor-dominated fluids eroded vent fillings and fissure walls within the dome and erupted fine-grained clastic material.

Thin, corridor-like tephra deposits that consist of fine-grained, hydrothermally altered dome rock and are associated with laharic breccias of the same composition may be used to identify periods of explosive hydrothermal activity within domes. Roobol and Smith (1975) have identified such deposits at Mont Pelée, Martinique, where they are interbedded with pumice deposits.

ROLE OF WATER IN FORMATION OF DOME-RELATED TEPHRA

In reviewing studies of explosive eruptions at domes, we have noted in all cases the role of water (liquid and vapor). Although this observation is not unusual in any explosive eruption, we believe that among other physical and chemical variables, the presence and migration of magmatic and meteoric water within the dome before, during, and after an eruptive epoch has

TABLE 4. TEPHRA GRAIN COUNTS OF LA SOUFRIÈRE DE GUADELOUPE
1976 ERUPTION (>38 μm FRACTION)

Component (%)	12 Aug Col Echelle	26 Aug Summit	30 Aug Savanne a Mulet	15 Sept Saint Claude
Plagioclase, unaltered	10.2	15.8	14.9	3.6
Plagioclase, unaltered	4.3	13.3	27.4	2.3
Orthopyroxene, unaltered	4.0	5.3	2.3	3.3
Orthopyroxene, altered	0.9	4.0	---	0.7
Clinopyroxene	0.3	---	0.6	1.3
Pyrite	0.6	2.2	2.9	0.7
Lithic fragments, unaltered	1.9	1.5	1.1	2.0
Lithic fragments, altered	75.8	56.7	45.7	84.4
Colorless glass, altered	0.9	0.3	0.6	---
Brown glass	0.9	0.3	0.6	---
Other	---	---	---	0.6
Total grains counted	322	323	163	307
Sulfur content*	1.7	2.7	4.3	1.3

*Sulfur determined by E. Gladney and D. Curtis, Los Alamos National Laboratory.

profound effects upon many of the tephra characteristics such as size, shape, and pyroclast surface chemical composition.

Migration of water (aqueous solutions) in dome lavas and its effect upon explosive phenomena and tephra compositions are reviewed in the following section. Migration may occur by a number of different physical processes, including flow along fractures, hydraulic injection from vent walls, molecular diffusion, and mechanical mixing of ground water into rising magma.

Where dome lavas are extruded, decompression and cooling cause diffusion of aqueous species from the lava. The high-temperature fluid that collects in fractures and voids within the lava is also at a high pressure when compared with the atmosphere. The dome rock strength (10 to 100 MPa depending upon composition and fracture density) can balance this overpressure; however, with continued movement and shifting of the dome lava, the mass displays metastable mechanical behavior. Some of the high-pressure steam may vent through fumaroles; this can relieve the overpressure, but can also cause a weakening of the lava by vapor-phase alteration. Continued fracturing, seismic pulses, or renewed heating of the dome by intrusion of a new magma pulse may cause the lava to catastrophically fail, leading to flashing of the confined pore water with subsequent dome explosion. Chemical evidence of this process is shown by vapor-phase mineral deposition in lithophysae and along fracture and foliation planes.

A different type of explosive behavior at domes is a consequence of the extruding dome lava coming into contact with meteoric water in the vent system. Under some circumstances, this contact will cause sudden chilling of the lava, which leads to formation of a network of fractures and lava granulation. In the event that the fragmented lava is more permeable than surrounding country rock, high pore pressures will drive water into the lava along fractures and over grain surfaces. Delaney (1982) has shown evidence of this mechanism along basaltic dike margins.

As a first approximation, these forms of dome water migration can be mathematically expressed as diffusion-type processes where a solute species, water, is moving within silicic, near-solidus magma (models applying Fick's laws of diffusion). We consider two cases of diffusion, the first being that of "slow" molecular transport of one material through another, which has attracted much experimental attention (e.g., Shaw, 1974). The differential equation modeling this case is generally linear and can be solved analytically. In particular, migration of magmatic volatiles from the lava to fracture planes, vesicles, and lithophysae exemplifies this type of diffusion (Fig. 19A). The second case is more difficult to analyze, both mathematically and experimentally, because it involves nonlinear terms in the model differential equation and "fast" mass reaction rates. This case is that of mixing of dome lava with ground water in the vent during extrusion (Fig. 19b). Although the scale of these two cases of diffusion are greatly different, the mathematical treatment discussed below will, we hope, add some insight to field and laboratory studies of domes.

A linear differential equation for the slow case can be written (e.g., see, Freer, 1981; Crank, 1975) in one dimension as

$$\partial c/\partial t = D\,(\partial^2 c/\partial x^2) \tag{1}$$

where c, the aqueous species concentration at any time t, and position x, is dependent upon a constant diffusion coefficient D. In order to reach equilibrium, the aqueous species moves through the lava from a concentrated (supersaturated) region to a subsaturated region. Equation 1 is a model of this migration in which water is "accelerated" from its initial localized concentration in the lava toward achievement of uniform distribution. This model requires a mass flux that is linearly proportional to the local concentration gradient. Equation 1, like that of the heat flow equation (Carslaw and Jaeger, 1959), is parabolic and can be solved using the same technique. For a semiinfinite source couple, the solution for c is obtainable using error function (erf) tables for

$$c(x, t) = \frac{c_0}{2}\left[1 - \mathrm{erf}\left(\frac{x}{2\sqrt{Dt}}\right)\right]. \tag{2}$$

Thus the initial concentration of water in dome lava (c_0) can be used to find how much might accumulate in fractures and pores after a certain amount of time, especially where c_0 is supersaturated at ambient pressures and temperatures. Typically, diffusion constants have been determined from experimental measurements and for this type of system are in the range of 10^{-9} to 10^{-6} cm^2/s (e.g., Freer, 1981).

Because diffusing magmatic water is an aqueous solution of ionic species of Na, K, and other elements, migration of these substances through the lava may be modeled with the Einstein diffusion equation:

$$c = \sigma\,(RT/z^2 F^2 D), \tag{3}$$

where the concentration of the ionic species c, is determined by the electrical conductivity σ, the gas constant R, the temperature T, the ionic charge z, the Faraday constant F, and the diffusion constant for an isotopic "tracer" of the studied species D. This analysis has been found to work quite well for diffusion of Na in obsidian (Carron, 1968). Overall, diffusivity decreases with increasing ionic radius and charge, a result quantitatively shown by Whittaker and Muntus (1970) for basalt. Therefore, where water migration has contributed to explosive behavior at domes, some chemical alteration of the related tephra should be expected, as was discussed earlier for the Mount St. Helens example.

The fast type of diffusion has been less studied, because it behaves mathematically in a highly nonlinear fashion. Situations where effects of grain boundaries, pressure and temperature gradients, and surface tension are present lead to a diffusion coefficient that is spatially dependent upon position. Fick's second law can be expressed as a nonlinear differential:

$$\partial c/\partial t = (\partial D/\partial x)\,(\partial c/\partial x) + D\,(\partial^2 c/\partial x^2). \tag{4}$$

For fast diffusion, D may have values in the range of 10^{-2} to 10^4 cm^2/s, so that the characteristic time scale for this type of diffusion is several orders of magnitude smaller than that of molecular diffusion. In analysis of the origin of these increased values of D, one may examine pressure and temperature effects. Activation energies and volumes have an exponential effect upon D (e.g., Hofmann, 1980); however, Lapham and others (1984) did not find this relationship for rhyolite obsidian at 850°C and pressures up to 500 MPa. In seeking mathematical solutions to equation 4, values of D as a function of c are required from measured $c(x)$ values. Expressing the initial conditions of equation 4 in terms of one variable $\eta = x/t^{1/2}$, c then becomes a function of η only, and equation 4 may be written as an ordinary homogeneous differential equation:

$$-\eta/2\,(dc/d\eta) = d/d\eta\,[D\,(dc/d\eta)], \tag{5}$$

for which the solution may be found by integrating (Shewmon, 1963):

$$-\frac{1}{2}\int_0^{c'} \eta\, dc = D\left[\frac{dc}{d\eta}\right]_0^{c'}, \tag{6}$$

and substituting for η

$$-\frac{1}{2}\int_0^{c'} x\, dc = Dt\left[\frac{dc}{dx}\right]_0^{c'}. \tag{7}$$

In this equation, dx can be considered a function of the fractured lava particle size, which is manifested by tephra grain diameter. With respect to meteoric water migration into silicic magma, initial fragmentation of the magma by thermal stresses provides fractures along which water moves rapidly by thermal hydraulic potential (Delaney, 1982). The effective diffusion is then only a matter of fracture spacing and fragment particle size (the characteristic distance dx, over which diffusion must take place in short times). For tephra with grain diameters on the order of several micrometres, substantial diffusion of various species can occur rapidly so that the bulk composition of fragmented material is altered quickly (e.g., Panum Crater example). This process likely involves solution and redeposition, the effects of which are observable under high resolution microscopy. For example, solution and redeposition may produce rapid changes in $^{18}O/^{16}O$ ratios to a depth of 0.1 μm as was observed in plagioclase in hydrothermal experiments by Giletti and others (1978). At greater depths in the solid material, chemical exchange is much slower.

A more specific treatment of diffusion along fractures and grain surfaces is given by Shewmon (1963), who analyzed grain boundary effects. In this analysis, Fick's second law is expressed with two diffusion coefficients: D_b relating to grain boundary diffusion, and D_1 the lattice or "normal" diffusion constant. The effects of these two types of diffusion may be combined into one expression:

$$\frac{\partial c}{\partial t} = D_b\left(\frac{\partial^2 c}{\partial y^2}\right) + \left(\frac{2D_1}{\delta}\right)\left(\frac{\partial c}{\partial x}\right) \tag{8}$$

where x is one half the thickness of the fracture δ, along a grain boundary surface of length y. The solution to equation 8 must

DIFFUSION OF H_2O IN DOME MAGMA

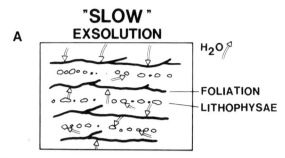

Figure 19. A: Sketch of the diffusion of aqueous species and water in silicic dome lavas. With decreasing pressure and temperature, diffusion concentrates fluids in vesicles, lithophysae, and along fracture and foliation planes. This "slow" diffusion process leads to metastable mechanical behavior of dome lavas. B: Sketch of meteoric waters "diffusing" into silicic magma from saturated country rock. This diffusion can be relatively fast because of surface area and grain-boundary effects caused by fracturing and large thermal gradients.

simultaneously satisfy the "normal" diffusion away from the grain boundary ($\partial c/\partial t = D_1 \nabla^2 c$). An example of this "double" diffusion (equation 8) is shown by Cole and others (1983), who modeled oxygen isotope exchange between fluid and rocks as a two-step process, involving first surface controlled reactions followed by diffusion through the solid material.

Finally, we consider the thermal diffusion effect, which has been named after Soret. It can be mathematically described and applied to systems that have a substantial thermal gradient. Here we subtract it from the flux term of Fick's first law:

$$J = -D \left(\frac{\partial c}{\partial x} \right) - \beta \left(\frac{dT}{dx} \right), \qquad (9)$$

where J is the solute flux, and β is the coefficient of solute flux describing the effect of the temperature gradient, dT/dx. In this situation, a lava dome intruded by new magma can develop a high temperature gradient and some species will respond by migrating either away or toward the higher temperature. Mineralization effects found in silicic domes (Burt and Sheridan, 1981) may be one example of this process, which has also been applied to melt differentiation in silicic magma chambers. A temperature gradient in a lava dome prior to renewed eruptive activity can contribute to accidental tephra showing strong chemical alteration.

In the preceding paragraphs we have discussed several diffusion processes that result in an apparent diffusion coefficient that varies greatly from that measured in laboratory experiments. These processes include: (1) "slow" molecular diffusion of aqueous species in near-solidus lava evidenced by lithophysae, vapor-phase crystallization and, in some cases, eventual dome explosion; (2) the effect of local water concentration, which can be analyzed as a function of grain size and is important where dome explosions have resulted from infiltration of ground water into the lava; (3) diffusion enhanced by the effects of grain surfaces in fragmented lava; (4) thermal gradient effects upon diffusion; and (5) the chemical alteration of dome lava by diffusion. In recognition of these various diffusion processes, Cole and others (1983) reported that the overall value of the diffusion coefficient D is fixed by grain size, shape (e.g., sphere, plate), and density, temperature, and water/solid mass ratio (specified rather than concentration c). Future microscopic studies of dome-related tephra may unfold the nature of nonlinear effects on water diffusion in domes.

In conclusion, diffusion of water in dome lavas encompasses a set of related processes that can lead to explosive eruptions. The relationship of diffusion to explosions is simply that it is the process by which volatile species, originally dissolved in the lava, are liberated for explosive expansion with tephra (phreatic and magmatic types). It also generally describes the mechanism of ground water mixing with extruding dome lava, which results in phreatomagmatic explosions. The importance of considering diffusion processes when studying tephra are that they are important underlying processes that contribute to observed tephra size distributions, grain shapes determined with an SEM, surface chemistry, and mechanism of dispersal (e.g., dry and wet surge, ballistic, and fallout).

EXPLOSIVE ERUPTIONS AT SILICIC DOMES AND CHARACTERISTICS OF THE TEPHRA DEPOSITS: SUMMARY

Four general types of explosive volcanic activity accompanying emplacement of silicic domes have been discussed: (1) initial Plinian and phreatomagmatic eruptions preceding dome emplacement; (2) Vulcanian explosions during continuing and episodic dome growth; (3) Peléean and Merapian activity associated with dome destruction; and (4) phreatic explosions preceding magmatic activity or accompanying energetic hydrothermal activity at older domes. Characteristics of the tephra, including volume of the deposit, bedding features, particle size, composition, and textures can be compared and contrasted when assessing explosive activity. These general characteristics are shown in Table 5 and are discussed below.

TABLE 5. DOME TEPHRA CHARACTERISTICS

Eruption Type	Deposits	Size	Tephra Composition	Texture
Initial Vent				
Magmatic	Plinian pumice fall and flow	Coarse (near vent) fall (0 to -3 ϕ)	Magma composition	Vesicular, angular
Phreatomagmatic	Dry surge	Fine ash (0 to 3 ϕ)	Slight to moderate surface alteration	Nonvesicular, slablike and abraded
Vulcanian	Wet and dry surge, coarse and fine fallout	Medium to fine ash (1 to 4 ϕ)	Fresh and altered juvenile, altered lithic clasts	Blocky and equant, nonvesicular to poorly vesicular, rounded
Peleean and Merapian	Poorly bedded avalanche and flow, bedded ash cloud surge	Blocks and ash (-5 to 1 ϕ)	Fresh magma and lithic clasts	Poorly vesicular and blocky
Phreatic	Poorly bedded, thin ash and lapilli mantles	Fine ash and minor lapilli (-1 to 3 ϕ)	Altered lithic clasts	Aggregated, complex shapes, "muddy"

Deposits

Bedding features and distribution are perhaps the most distinguishing field characteristic of dome-related pyroclastic deposits. Although deposits of past eruptions may only be partly preserved and exposed in the geologic record, one or more stratigraphic sections can provide ample data to characterize the activity.

Vent-forming eruptions preceding dome emplacement generally produce a sufficient tephra volume (0.05 to 1.0 km^3) that deposits near the vent may be up to 10 m or more in thickness; several centimeters of ash may be found over areas of 100 km^2 or more. Pumice beds and thinly bedded pyroclastic surge layers typify these deposits. The surge beds occur within several kilometers of the vent and are termed "dry" after Wohletz and Sheridan (1983) because they include abundant dune forms, are poorly to non-indurated, have bedding angles of less than 10°, and show little or no soft-sediment deformation features. In flat terrain, surge deposits have abundant dunes near vent and planar bed forms in distal sections (Wohletz and Sheridan, 1979).

Vulcanian eruptions associated with dome growth produce smaller tephra volumes of generally less than 0.1 km^3. These tephra form several centimeter- to several meter-thick mantles on the flanks of the volcano. Bedding is characterized by alternating layers (1 to 5 cm thick) of coarse- and fine-grained fallout, dry and wet surge deposits, and lahars near the base of the volcano. Where dry-surge deposits are found on cone flanks, a reversed facies distribution may be found (relative to distribution illustrated by Wohletz and Sheridan, 1979). Planar bedding is common near the vent and dune beds are abundant near the base of the cone. Wet-surge deposits (Wohletz and Sheridan, 1983) are characterized by bedding planes dipping up to 25°, and planar and massive beds, abundant accretionary lapilli, vesiculated tuff, soft-sediment deformation, induration, and strong tephra alteration. Many Vulcanian tephra deposits grade laterally away from the vent into lahars.

Dome and lava-lobe destruction by Peléean and Merapian explosions produce more chaotic tephra deposits, including poorly bedded and sorted pyroclastic flow and avalanche beds that are associated with well-sorted fine layers (ash-cloud surges) (Fisher, 1979; Fisher and Heiken, 1982). These deposits generally thicken downslope and fill drainage areas. Their volume is small (<1.0 km^3).

Phreatic tephra layers are typified by their thinness (<1 m and generally less than several centimeters) and extremely small volume (<0.0001 km^3). They are also poorly bedded ash and lapilli of muddy appearance and generally very fine grained, except for centimeter- and meter-sized clasts near the vent.

Tephra Characteristics

Particle size distributions, composition, and microscopic grain textures also characterize eruptive style. These features can be qualitatively assessed in the field but lend themselves to detailed quantitative laboratory study.

Peléean and Merapian tephra have the coarsest particles and, in most exposures, include large blocks. Vulcanian and phreatomagmatic tephra are fine to medium ash, and phreatic ash is very fine grained. Plinian tephra is coarse near the vent where exposures are most apparent (see Table 5). Analyses of grain size (such as those presented in Fig. 20) can illustrate information on the relative eruptive energy and dispersal mechanisms that are responsible for preferential sorting and development of size distributions.

Chemical compositions of tephra are rarely reported because of analytical difficulties; however, component analysis reveals important constituents including phenocryst, glass, accessory, and lithic clasts. These data allow interpretation of tephra origin (magma composition and physical properties, hydrothermally altered accessories, and lithic clast types) and emplacement mechanisms that can alter initial constituents. Initial Plinian eruptions distribute tephra that reflect the juvenile magma composition; tephra alteration is minor. Initial phreatomagmatic tephra can be strongly altered; particles have compositions that vary

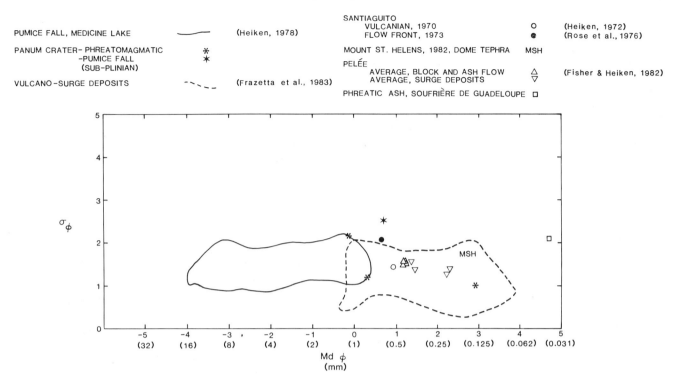

Figure 20. Sorting coefficient (σ_ϕ) vs. median diameter (Md_ϕ) plot summarizing size characteristics of dome-related tephra discussed in this paper.

from fresh juvenile composition to hydrated glass or secondary minerals (see Panum Crater description). Vulcanian tephra have both juvenile magma and lithic components that have altered surface compositions (as described for Mount St. Helens). Peléean and Merapian tephra compositions also display lithic and fresh magma components, but the amount of clast alteration is generally small, except in cases where the dome has been extensively hydrothermally altered. Finally, phreatic ash contains no juvenile clasts and lithic clasts are extensively hydrothermally altered to clays, zeolites, and some sulfides (see section on La Soufrière de Guadeloupe).

Tephra grain textures observed with the SEM have been emphasized in this paper because they show many characteristics, of which only the most general are listed in Table 5. In addition to readily observable textures such as vesicularity and grain shape, the variation among surface features, such as abrasion scratches, chips, pits, alteration coatings, adhering materials, grain rounding, fracture surfaces, and aggregations provides a complex data set for characterization. This work, still in progress, will require further development of SEM techniques for quantification.

ACKNOWLEDGMENTS

We thank Jonathon Fink, the organizer of this symposium, who encouraged us to pull the bits and pieces together for this paper. The manuscript made a lot more sense after thorough reviews by Stephen Self and Richard V. Fisher. Much of the discussion of the tephra formation in phreatic eruptions was part of an unpublished document done many years ago in collaboration with Michel Semet and Michel Feuillard. Discussions with Marty Horn were also very helpful.

REFERENCES CITED

Burt, D. M., and Sheridan, M. F., 1981, Model for the formation of uranium/lithophile element deposits in fluorine-rich volcanic rocks: American Association of Petroleum Geologists Studies in Geology 13, p. 99–109.

Carron, J-P., 1968, Auto diffusion du sodium et conductivitie électrique dans les obsidiènnes granitiques: Paris Académie de Science, Comptes Rendus, ser. D, v. 266, p. 854–856.

Carslaw, H. S., and Jaeger, J. C., 1959, Conduction of heat in solids, Second Edition: London, Oxford University Press, 510 p.

Cashman, K. V., and Taggart, J. E., 1983, Petrologic monitoring of 1981 and 1982 eruptive products from Mount St. Helens: Science, v. 221, p. 1385–1387.

Crank, J., 1975, The mathematics of diffusion: Oxford, Oxford University Press, 414 p.

Cole, D. R., Ohmoto, H., and Lasaga, A. C., 1983, Isotope exchange in mineral-fluid systems. I. Theoretical evaluation of oxygen isotopic exchange accompanying surface reactions and diffusion: Geochimica et Cosmochimica Acta, v. 47, p. 1681–1693.

Delaney, P. T., 1982, Rapid intrusion of magma into wet rock: Groundwater flow due to pore-pressure increases: Journal of Geophysical Research, v. 87, p. 7739–7756.

Fink, J., 1983, Structure and emplacement of a rhyolitic obsidian flow: Little Glass Mountain, Medicine Lake Highland, northern California: Geological Society of America Bulletin, v. 94, p. 362–380.

Fisher, R. V., 1979, Models for pyroclastic surges and pyroclastic flows: Journal of Volcanology and Geothermal Research, v. 6, p. 305–318.

Fisher, R. V., and Heiken, G. H., 1982, Mt. Pelée, Martinique: May 8 and 20, 1902, pyroclastic flows and surges: Journal of Volcanology and Geothermal Research, v. 13, p. 339–371.

Fisher, R. V., Smith, A. L., and Roobol, M. J., 1980, Destruction of St. Pierre, Martinique by ash cloud surges, May 8 and 20, 1902: Geology, v. 8, p. 472–476.

Frazzetta, G., Gillot, P. V., LaVolpe, L., and Sheridan, M. F., 1984, Volcanic hazards at Fossa of Vulcano: Data from the last 6,000 years: Bulletin Volcanologique, v. 47, p. 105–124.

Frazzetta, G., La Volpe, L., and Sheridan, M. F., 1983, Evolution of the Fossa cone, Vulcano, *in* Sheridan, M. F., and Barberi, F., eds., Explosive volcanism: Journal of Volcanology and Geothermal Research, v. 17, p. 329–360.

Fisher, R. V. and Schmincke, H.-U., 1984, Pyroclastic rocks: Berlin, Springer-Verlag, 472 p.

Freer, R., 1981, Diffusion in silicate minerals and glasses: A data digest and guide to the literature: Contributions to Mineralogy and Petrology, v. 76, p. 440–454.

Giletti, B. J., Semet, M. P., and Yund, R. A., 1978, Studies in diffusion—II. Oxygen in feldspars: An ion microprobe determination: Geochimica et Cosmochimica Acta, v. 42, p. 45–57.

Healy, J., 1976, Geothermal fields in zones of recent volcanism: Proceedings of the U.N. Second Conference on Geothermal Energy, Lawrence Berkeley Laboratory, Berkeley, p. 415–422.

Heiken, G. H., 1972, Morphology and Petrography of volcanic ashes, Geological Society of America Bulletin, v. 83, p. 1961–1983.

Heiken, G., 1978, Plinian-type eruptions in the Medicine Lake Highland, California, and the nature of the underlying magma: Journal of Volcanology and Geothermal Research, v. 4, p. 375–402.

Heiken, G., and Eichelberger, J., 1980, Eruptions at Chaos Crags, Lassen Volcanic National Park, California: Journal of Volcanology and Geothermal Research, v. 7, p. 443–481.

Heiken, G., and Wohletz, K., 1985, Volcanic ash: Berkeley, University of California Press, 264 p.

Heiken, G., Crowe, B., McGetchin, T., West, F., Eichelberger, J., Bartram, D., Peterson, R., and Wohletz, K., 1980, Phreatic eruption clouds: The activity of La Soufrière de Guadeloupe, F.W.I., August-October, 1976: Bulletin Volcanologique, v. 43, p. 383–395.

Hofmann, A. W., 1980, Diffusion in natural silicate melts: A critical review, *in* Hargraves, R. B., ed., Physics of magmatic processes: Princeton, New Jersey, Princeton University Press, p. 387–417.

Hopson, C. A., and Melson, W. G., 1984, Eruption cycles and plug-domes at Mount St. Helens: Geological Society of America Abstracts with Programs, v. 16, p. 544.

Kistler, R. W., 1966, Geologic map of the Mono Craters quadrangle, Mono and Tuolomne counties, California: Geological Survey Quadrangle Map GQ-462, scale 1:62,500.

LaCroix, A., 1904, La Montagne Pelée et ses Éruptions: Paris, Masson et Cie, 662 p.

Lapham, K. E., Holloway, J. R., and Delaney, J. R., 1984, Diffusion of H_2O and D_2O in obsidian at elevated temperatures and pressures: Journal of Non-Crystalline Solids, v. 67, p. 179–191.

Lasserre, G., 1961, La Guadeloupe: Paris, Union Francais d'Impression, 1135 p.

Melson, W. G., 1983, Monitoring the 1980-1982 eruptions of Mount St. Helens: Compositions and abundances of glass: Science, v. 221, p. 1387–1391.

Mercalli, G., and Silvestri, D., 1891, Le eruzioni dell'isola di Vulcano, incominciate il 3 Agosto 1888 e terminate il 22 Marzo 1890: Relazione Scientifica, 1891, Annuale Ufficiale Centrale Meteorologia e Geodintorno, v. 10, no. 4, p. 1–213.

Neumann van Padang, 1931, Der Ausbruch des Merapi (Mittel Java) im Jahre 1930: Zeitschrift für Vulkanologie, v. 14, p. 135–148.

Newhall, C. G., and Melson, W. G., 1983, Explosive activity associated with the growth of volcanic domes: Journal of Volcanology and Geothermal Research, v. 17, p. 111–131.

Putnam, W. L., 1938, The Mono Craters, California: Geographic Review, v. 28, p. 68–82.

Rittmann, A., 1936, Vulkane und ihre Tätigkeit: Stuttgart, Ferdinand Enke, Stuttgart, 188 p.

Roobol, M., and Smith, A., 1975, A comparison of the recent eruptions of Mt. Pelée, Martinique and Soufriere, St. Vincent: Bulletin Volcanologique, v. 39, p. 214–240.

Rose, W. I., Jr., 1973, Pattern and mechanism of activity at the Santiaguito volcanic dome, Guatemala: Bulletin Volcanologique, v. 37, p. 73–94.

Rose, W. I., Jr., Pearson, T., and Bonis, S., 1976, Nuée ardente eruption from the foot of a dacite lava flow, Santiaguito Volcano, Guatemala: Bulletin Volcanologique, v. 40, p. 23–28.

Shaw, H. R., 1974, Diffusion of H_2O in granitic liquids: Part I. Experimental data; Part II. Mass transfer in magma chambers, *in* Hofmann, A. W., Giletti, B. J., Yoder, H. S., and Yund, R. A., eds., Geochemical transport and kinetics: Washington, D.C., Carnegie Institution of Washington Publication 634, p. 139–170.

Sheridan, M. F., Moyer, T. C., and Wohletz, K. H., 1981, Preliminary report on the pyroclastic products of Vulcano: Memorandum Sociale Astronomia Italia, v. 52, p. 523–527.

Shewmon, P. G., 1963, Diffusion in solids: New York, McGraw-Hill, 203 p.

Sieh, K., 1983, Most recent eruption of the Mono Craters, eastern central California: EOS (American Geophysical Union Transactions), v. 64, no. 8, p. 889.

Swanson, D. A., Casadevall, T. J., Dzurisin, D., Malone, S. D., Newhall, C. G., and Weaver, C. S., 1983, Predicting eruptions at Mount St. Helens, June 1980 through December 1982: Science, v. 221, p. 1369–1376.

Waitt, R. B., Jr., Pierson, T. C., MacLeod, N. S., Janda, R. J., Voight, B., and Holcomb, R. T., 1983, Eruption-triggered avalanche, flood, and lahar at Mount St. Helens—Effects of winter snowpack: Science, v. 221, p. 1394–1397.

Walker, G., 1971, Grain-size characteristics of pyroclastic deposits: Journal of Geology, v. 79, p. 696–714.

Westercamp, D., and Traineu, H., 1983, The past 5,000 years of volcanic activity at Mt. Pelée, Martinique (F.W.I.): Implications for assessment of volcanic hazards: Journal of Volcanology and Geothermal Research, v. 17, p. 159–185.

Whittaker, E.J.W., and Muntus, R., 1970, Ionic radii for use in geochemistry: Geochimica et Cosmochimica Acta, v. 34, p. 945–956.

Wohletz, K., and Crowe, B., 1978, Development of lahars during the 1976 activity of La Soufrière de Guadeloupe: Geological Society of America Abstracts with Programs, v. 10, p. 154.

Wohletz, K. H., and Sheridan, M. F., 1979, A model of pyroclastic surge, *in* Chapin, C. E., and Elston, W. E., eds., Ash-flow tuffs: Geological Society of America Special Paper 180, p. 177–194.

——, 1983, Hydrovolcanic explosions II: Evolution of basaltic tuff rings and tuff cones: American Journal of Science, v. 283, p. 385–413.

Wohletz, K., and Krinsley, D., 1982, Scanning electron microscopy of basaltic hydromagmatic ash: Los Alamos National Laboratory document LA-UR 82-1433, 29 p.

Wood, S. H., 1977, Distribution, correlation, and radiocarbon dating of late Holocene tephra, Mono and Inyo Craters, eastern California: Geological Society of America Bulletin, v. 88, p. 89–95.

MANUSCRIPT ACCEPTED BY THE SOCIETY MAY 5, 1986

Origin of pumiceous and glassy textures in rhyolite flows and domes

Jonathan H. Fink
Curtis R. Manley
Geology Department
Arizona State University
Tempe, Arizona

ABSTRACT

Surface mapping and microscopic observation of textures in glassy and pumiceous rocks from several groups of silicic lava flows and domes, along with drill cores of the Inyo Scientific Drilling Project, indicate that development of the textural stratigraphy of the flows is controlled by a combination of cooling, microfracturing, and migration of gases released by crystallization. The Inyo cores have provided near-vent and distal views of the interior of the 550-yr-old Obsidian Dome rhyolite flow, as well as a profile through the unerupted portion of its feeder dike. The flow stratigraphy revealed in the drill core and in the fronts of several other Holocene-age silicic flows consists of a finely vesicular pumice carapace underlain successively by obsidian, coarsely vesicular pumice, obsidian with lithophysae, crystalline rhyolite, more obsidian with lithophysae, and basal breccia.

The obsidian layers form where rapid cooling inhibits diffusion of ions and prevents crystallization. The transition from surface pumice to obsidian is controlled by the depth at which overburden pressure suppresses vesiculation. The thickness of the rigid crust is determined by the rapid decrease in the lava's temperature-dependent yield strength with depth. The coarsely vesicular pumice layer forms as gases released by crystallization rise through microcracks and are trapped beneath the rigid pumiceous surface layer. Thickness and buoyancy of the coarsely vesicular pumice layer increase with flow length, eventually giving rise to diapirs that rise to the surface of the largest flows. Increasing gas content of the coarsely vesicular pumice layer in active flows can also lead to such volcanic hazards as explosive craters on distal flow surfaces or pyroclastic flows triggered by collapse of flow fronts.

INTRODUCTION

The hazards associated with erupting silicic magmas depend in large part on their water contents. Most silicic magma bodies are believed to be stratified with respect to volatiles (Kennedy, 1955; Smith, 1979; Hildreth, 1981; Blake, 1981; Sparks and others, 1984). Consequently, eruptions frequently tap magmas of progressively decreasing water content, resulting in the commonly observed sequence of explosive activity followed by quieter emplacement of silicic lava flows and domes. Field evidence of this progression is seen in the widespread occurrence of silicic domes within tephra rings, or silicic lava flows overlying contemporaneous pyroclastic deposits of nearly identical chemical composition.

Recent studies by Eichelberger and associates (e.g., Eichelberger and Westrich, 1981; Taylor and others, 1983) have documented decreases in juvenile water content that accompany the above-mentioned transitions from explosive to effusive activity. Sequences of tephra layers beneath silicic domes and flows at Medicine Lake Highland Volcano, Newberry caldera, Long Valley caldera, and Lassen Peak all show progressive upward decreases in water content from around 2 wt% to about 0.5 wt%, whereas the overlying lava flows have between 0.1 and 0.2 wt%.

Silicic lava flows are characterized by a wide range of surface textures that vary in vesicularity, crystallinity, color, and flow layering. The origin of these differences has generally been attributed to local chemical inhomogeneities, particularly with respect to volatile content. If the relationships between volatile content and texture could be clearly delineated, then studies of lava flows could better supplement those of pyroclastic deposits in providing useful constraints for assessment of hazards.

Initial attempts to extend volatile measurements from tephra deposits to the various pumiceous and glassy textures found in silicic domes and flows were hampered by difficulties in obtaining interior samples from extrusions young enough to preserve glass (B. Taylor, 1985, personal commun.). Now, however, drill cores of the Inyo and Valles Scientific Drilling Projects (Eichelberger and others, 1984; Goff and others, 1984) provide three nearly complete cross sections through glassy rhyolite flows in California and New Mexico. Accurate thickness measurements coupled with the excellent sampling opportunities offered by the cores allow the textural variations in rhyolite flows to be correlated with water content, stratigraphic position, and microstructure.

In this paper we first describe the distribution of textures seen in rocks on the upper surfaces and fronts of rhyolite flows and domes, then present macroscopic and scanning electron microscopic views and water content data for the Inyo drill core samples. We then propose a model for the role of volatiles in the development of pumiceous and glassy textures during the emplacement of rhyolite flows and domes, and finally discuss the model's implications for volcanic hazards associated with these extrusions.

OBSERVATIONS OF TEXTURES

Field Observations

The various textures observed in rhyolite flows reflect both the volatile content and emplacement conditions of the magma. Before attempting to correlate these textural units with the conditions of their formation, we need to first describe the geometric relationships among the units as revealed on flow surfaces and by the Inyo drill cores.

Figure 1 is an aerial photograph of Little Glass Mountain, a relatively voluminous 1100-yr-old rhyolitic obsidian flow on the Medicine Lake Highland Volcano in northern California (Anderson, 1941; Heiken, 1978; Johnston and Donnelly-Nolan, 1981; Fink, 1983). Rocks on the surface of this flow show three principal textures, two of which appear as light and dark areas on the photograph. The light-colored material is a nearly aphyric, finely vesicular pumice (FVP) which forms a carapace covering much of the flow surface. The dark lava is primarily coarsely vesicular pumice (CVP) which rises from the flow interior to the surface, especially near the distal margins. Glassy obsidian (OBS), the third unit, forms narrow dark zones separating the CVP and FVP units. Individual outcrops of CVP have domal structures and, on some lobes, are distributed with regular spacings in the downstream direction (Fig. 1). In flow fronts, the CVP is seen to form a more porous, less-dense layer beneath the FVP and OBS. The FVP and OBS layers are each typically between 5 and 10 m thick in flow fronts. These observations are consistent with a model (Fink, 1980, 1983) in which the buoyant CVP rises to the surface through the overlying OBS and FVP layers as a result of a Taylor or gravity instability, leading to regularly spaced diapiric domes of CVP on the upper-flow surface. These same textural relationships are also observed on dacite flows such as the Medicine Lake glass flow near Little Glass Mountain.

Another key observation about the CVP unit comes from comparisons of its distribution on different domes believed to be fed by the same magma body. The two vent areas on Little Glass Mountain align with a series of other rhyolite domes and flows and an extensive set of ground cracks and depressions to the northeast (Fig. 2a). These structural features have been interpreted as evidence for the presence of an underlying silicic dike or dike set that fed the extrusions (Fink and Pollard, 1983). When the quantities of CVP exposed on the surfaces of the different domes along this trend are compared, there appears to be a disproportionate amount in the larger extrusions. Figure 2b is a graph showing the areas of CVP outcrops on the domes along the Little Glass Mountain trend versus the total areas of those domes. The largest domes have a relatively great amount of CVP exposed, whereas the small domes have none. Similar relationships can be observed among other sets of Holocene rhyolite domes and flows such as those on South Sister volcano in Oregon (Williams, 1944; Scott, this volume; Figs. 3a,b) and the Inyo domes (Miller, 1985; Figs. 4a,b) in eastern California. Apparently, the size of an extrusion is a decisive factor in determining whether or not it displays the CVP texture.

Explosion craters are occasionally found on the surfaces of rhyolite flows. On Little Glass Mountain and on the neighboring Glass Mountain flow, these pits range up to 50 m in diameter and 15 m in depth (Eichelberger, 1974), and are commonly lined with obsidian containing anomalously large vesicles (up to 5 cm in diameter). Ejected blocks up to 0.5 m in diameter can be recognized up to 100 m away from these craters, reflecting their explosive origin. Even though these craters extend down as far as one-third of the depth of the flow, they reveal the same three principal textures seen in the flow fronts and on the upper surfaces. Similarly, some well-dissected Miocene age rhyolite flows in Idaho exhibit large gas cavities (up to several metres across) beneath an upper glassy zone and above a crystalline interior (Bonnichsen, 1982).

Inyo Drill Core Observations

Several of the more puzzling aspects of the Inyo drill core involve the stratigraphic relations among the different textures, and in particular the CVP unit. In flow fronts it is not possible to determine the thickness of this layer because its base is obscured by talus slopes. In the two drill cores through the Inyo Obsidian Dome flow, the CVP occurs as a 5–10-m-thick layer sandwiched between upper and lower obsidian zones. The upper obsidian forms an aphyric, flow-banded, glassy, black layer beneath the white to gray FVP crust. The lower obsidian grades downward through a series of lithophysal zones to a fully crystalline, dis-

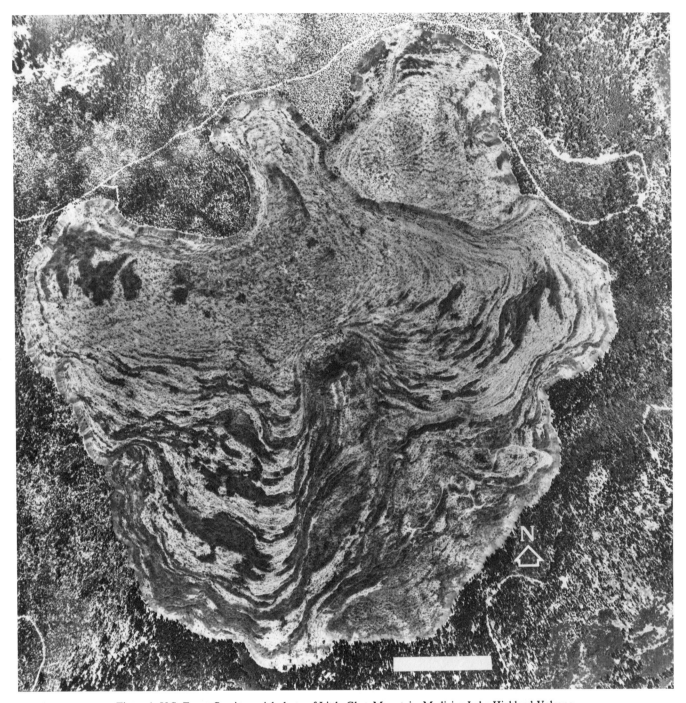

Figure 1. U.S. Forest Service aerial photo of Little Glass Mountain, Medicine Lake Highland Volcano, northern California. Areas of diapirically emplaced coarsely vesicular pumice show up as dark features contorted by subsequent flow of the lava. Scale bar = 500 m.

tinctly flow-banded rhyolite interior. Beneath the central zone, the sequence is reversed, proceeding from the crystalline rhyolite through lithophysal obsidian to glassy obsidian, and finally passing down to the basal breccia zone (Fig. 5).

The attitudes of flow banding in a silicic dome or flow reflect how the lava moved during emplacement. Dissected flows commonly contain a zone of subhorizontal foliations extending from the base to within several metres of the upper surface. Near the vent areas of these same extrusions, flow layering is subvertical throughout the entire cross section. These relations have been interpreted (e.g., Fink, 1983) as being the combined results of flow through a conduit and along the ground. As magma rises

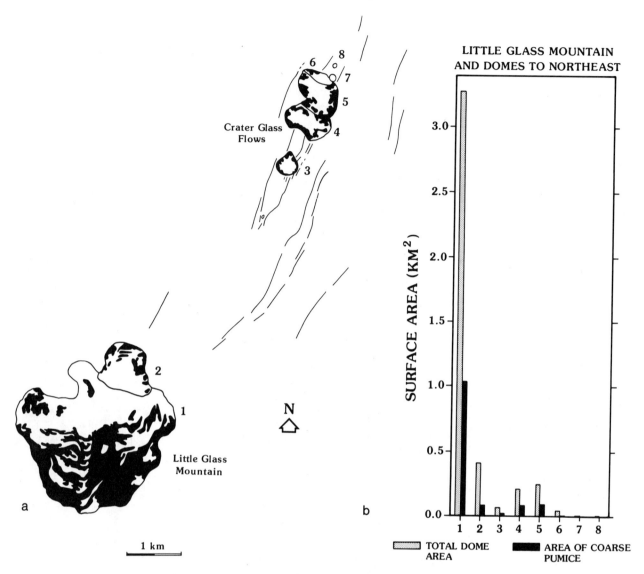

Figure 2. Medicine Lake Highland Volcano, northern California. a: Sketch map showing outlines of the 1100-yr-old rhyolite domes and flows on the west side of the Medicine Lake Highland Volcano, as well as ground cracks, faults, and coarsely vesicular pumice (dark areas). Little Glass Mountain is the southernmost flow. Numbers 1-9 are sites of rhyolite extrusions. b: Bar graph showing areas of coarsely vesicular pumice outcrops and total areas for the rhyolite extrusions shown in Figure 2a, numbered south to north.

through a conduit, it develops vertical foliation in response to shear stresses along the walls. Emerging lava rises vertically above the vent, rifts, and then spreads laterally, inheriting nearly vertical layering. As the lava moves outward, shear stresses at the flow base can cause the flow layering to rotate to horizontal and, as a consequence, a zone of horizontal layers propagates upward through the flow as it advances. This rotation can only occur in those parts of the flow that behave plastically. The cooler, brittle surface crust retains it original vertical foliation as it is rafted passively along by the flowing material beneath.

The Inyo drill cores through Obsidian Dome both reveal predominantly horizontal flow layering; only the upper part of the flow shows vertical foliation (Fig. 5). The thickness of the vertically foliated zone varies from about 20 m near the flow margin to about 16 m near the vent. In both cores, the transition from horizontal to vertical layering occurs at the top of the CVP layer. Thus, the CVP appears to be located immediately beneath the brittle, rigid crust of the flow.

Another distinctive attribute of the CVP layer is its anomalously high volatile content. Measurements of juvenile water content by Westrich and Eichelberger (1984) showed concentrations of between 0.1 and 0.2 wt% throughout most of both drill cores.

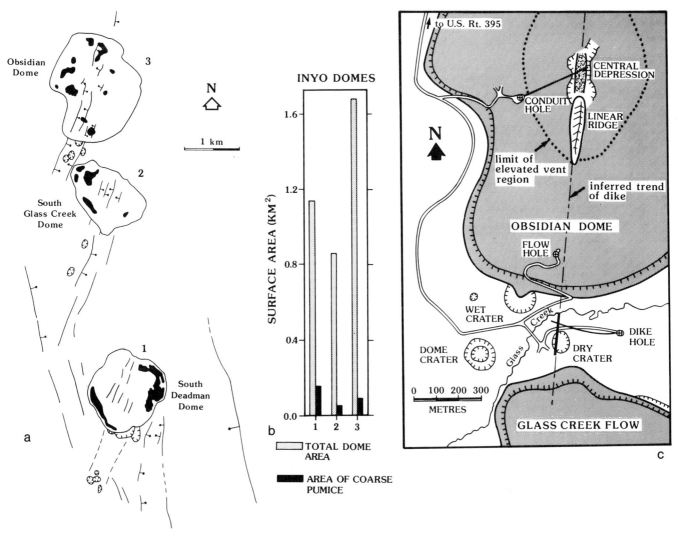

Figure 3. Inyo Domes, eastern California. a: Sketch map of the three major 600-yr-old Inyo rhyolite domes and their vicinity, showing areas of coarsely vesicular pumice (dark areas), normal faults, ground cracks (dashed where inferred), and phreatic pits. b: Bar graph showing areas of coarsely vesicular pumice outcrops and total areas for the domes shown in Figure 3a, numbered south to north. c: Map of the southern portion of Obsidian Dome and vicinity showing the locations of the Inyo Scientific Drilling Project holes: Flow Hole (RDO-2A), Conduit Hole (RDO-2B), and Dike Hole (RDO-3A). The latter two are slant holes. The names of phreatic craters are informal. Modified from Eichelberger and others (1984).

The exception in both cases was the CVP layer, which had between 0.4 and 0.6 wt% (Fig. 5). Isotopic studies (Taylor and others, 1983; O'Neil and Taylor, 1985) have demonstrated that this water is magmatic, rather than meteoric. These relations suggest that either the CVP layer has been enriched in volatiles, or the rest of the flow has been depleted, or some combination of the two.

The distribution of flow layering within the drill core can also help constrain the timing of crystallization. The central crystalline rhyolite zone is pervasively foliated, whereas the surrounding obsidian contains individual lithophysae that become more flattened and distorted toward the flow interior. The crystalline zone and deformed lithophysae thus formed before flow had ceased, and the spherical, undistorted lithophysae developed after the flow stopped moving.

Microscopic Observations

Several important observations of the flow textures require microscopic examination. We have used backscatter (BSE) and secondary electron (SE) modes of scanning electron microscopy (SEM) to look at the distribution of vesicles, crystals, cracks, and flow banding in samples from the two Obsidian Dome drill cores

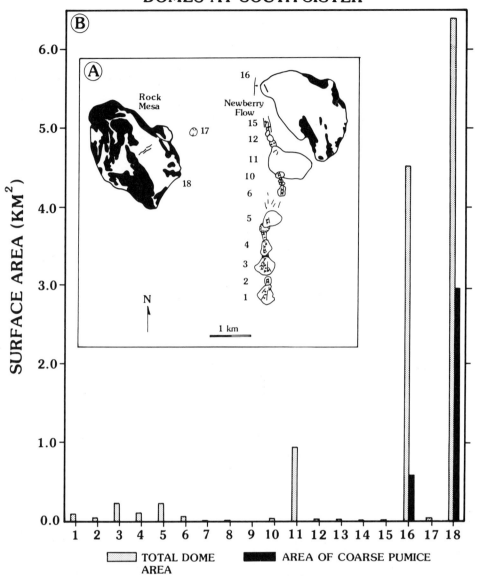

Figure 4. South Sister Volcano, central Oregon. a: Sketch map of the Holocene rhyolite extrusions (numbers 1–18) on the south flank of the volcano, showing faults, ground cracks, and areas of coarsely vesicular pumice (dark areas) exposed on the dome surfaces. b: Bar graph showing areas of coarsely vesicular pumice outcrops and total areas for rhyolite flows shown in Figure 4a. Domes along the (eastern) Devil's Hill chain are numbered south to north. Numbers 18 and 17 are Rock Mesa and the adjacent small dome, respectively.

(Fink and others, 1985). We will present some of these observations here, proceeding from the crystalline center of the flow toward the upper surface.

The crystalline rhyolite shows well-developed flow foliation consisting of alternating submillimetre-scale layers of crystal-rich and crystal-free lava (Fig. 6a). The crystal-rich bands contain quartz and feldspar crystals as well as numerous angular vesicles which are absent from the aphyric bands. The vesicles constitute up to 20% of the volume of the layers. Figure 6a also shows a few vertically oriented microfractures that cross the bands. These are common throughout the center of the flow, and extend as far upward as the CVP layer. Within the CVP layer itself, microfractures commonly connect large vesicles. Some of the larger lithophysae have grown over these cracks, again suggesting that they represent crystallization that occurred after the flow came to rest. The uppermost OBS and FVP zones show few of these microcracks (Fig. 6b), although the FVP is broken by much larger vertical joints that extend from the upper surface.

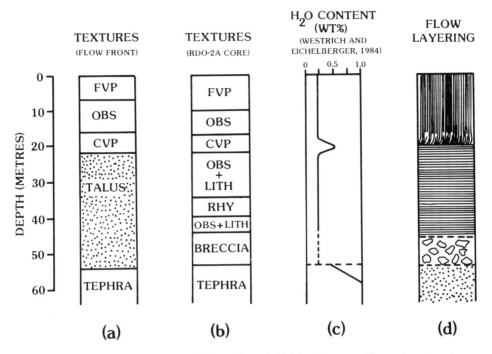

Figure 5. Cross sections through the distal portion of Obsidian Dome. a: Textural stratigraphy as observed from the flow front. Talus obscures lower portions. b: Stratigraphy as revealed in drill hole RDO-2A. FVP = finely vesicular pumice; OBS = obsidian; CVP = coarsely vesicular pumice; LITH = lithophysae; RHY = crystalline rhyolite. c: Schematic profile of water contents of fresh glass samples measured by Westrich and Eichelberger (1984). Note the anomaly associated with the CVP layer. d: Schematic representation of flow-banding attitudes in RDO-2A core.

BSE and SE images of the CVP layer suggest that many of the vesicles had a complex history that may have included alternating periods of inflation and collapse. Many vesicles in the CVP zone are contorted and flattened, in some cases resulting in vesicles that show transitions to horizontal cracks (Fig. 6c). As seen in Figure 6d, many of the CVP vesicles are filled with debris that is identical in appearance to the glass found in the vesicle walls. Close inspection of these walls reveals additional fractures parallel to the vesicle margins. The fragments are not equally distributed within individual vesicles, but tend to be concentrated in certain areas and absent from others. The vacant zones are commonly circular in outline and tend to be located near sites where fractures connect the vesicles.

The FVP and OBS layers appear similar except for the presence of small vesicles in the FVP (Fig. 6b). Microphenocrysts like those found in the central crystalline zone are nearly absent, and flow foliation is subvertical and poorly developed. Larger phenocrysts in the FVP tend to be euhedral, in contrast to the subhedral to anhedral shapes of phenocrysts in the lower portions of the flow.

MODEL FOR DEVELOPMENT OF TEXTURES IN RHYOLITE FLOWS

The large- and small-scale observations presented above provide several constraints on how textures can develop during the emplacement of rhyolite flows and domes. Although various models have been previously proposed, none were based on the full range of data described here. In this section we present our interpretation of the above observations along with some relevant calculations of stresses within lava flows. Because many of the mechanisms involved in formation of the textures are interrelated and occur simultaneously, a strict chronological or stratigraphic presentation is not possible. In this section we will first give an overview of our model, then discuss some relevant rheologic studies, and then address the formation of specific textures.

Our model includes three principal steps: (1) crystallization within the flow releases dissolved magmatic volatiles; (2) advance of the lava forms microcracks through which these gases can move upward; and (3) cooling of the upper surface increases the yield strength and creates a nondeforming crust through which the rising gases are unable to migrate (Fig. 7). As a result of these three steps, a gas-rich, low-density layer capable of rising buoyantly to the flow surface develops beneath the brittle crust. Crystallization appears to be controlled by the ability of ions to migrate and thus occurs preferentially in the central portion of the flow where maintenance of higher temperatures and lower viscosities allows more rapid diffusion. Evaluation of this model is hindered by a lack of knowledge of some of the physical properties of rhyolite magmas.

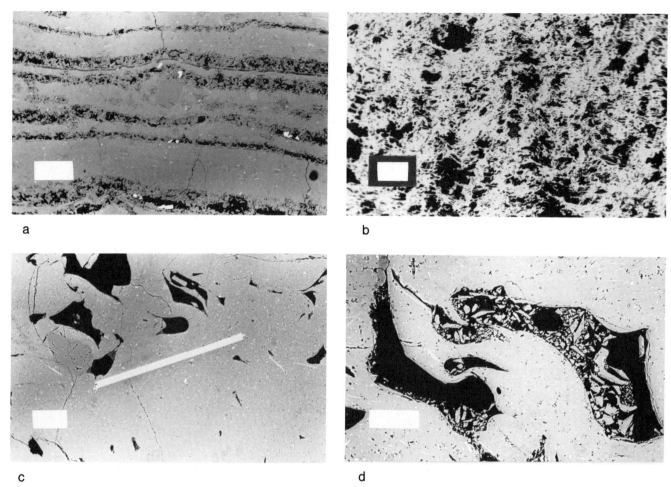

Figure 6. Backscatter mode scanning electron micrographs of samples from RDO-2A drill core. Scale bars below each photograph. a: Crystalline rhyolite from a depth of 36.9 m showing interlayered crystal-rich and crystal-poor bands. Note the relatively high concentration of angular vesicles in those layers that underwent crystallization. Vesicles inferred to represent volatiles released by crystallization. Scale bar = 0.5 mm. b: Finely vesicular pumice from depth of 1.7 m showing angular shapes of vesicles, absence of phenocrysts, and lack of throughgoing microfractures. Scale bar = 0.5 mm. c: Coarsely vesicular pumice from depth of 20.1 m. Note contorted shapes and smooth walls of vesicles, and numerous fractures connecting vesicles. Scale bar = 0.2 mm. d: Coarsely vesicular pumice from depth of 20.1 m showing inclusions spalled from smooth walls of larger vesicles. Scale bar = 0.1 mm.

Rhyolite lavas have been modeled as temperature-dependent, non-Newtonian materials with finite yield and tensile strengths (e.g., Shaw, 1965; Murase and McBirney, 1973; Hulme, 1974; Fink, 1983; Spera and others, 1982; McBirney and Murase, 1984). When the applied shear stresses exceed the yield strength, the lava is capable of flowing, and when the stresses exceed the tensile strength, the lava can fracture. By analogy with viscosity, both types of strength are assumed to decrease with increasing volatile content and temperature. Hence in a cooling flow, the temperature dependence of rheology will lead to inward decreases in yield strength, tensile strength, and viscosity.

Laboratory data on the yield and tensile strengths of rhyolitic lavas are quite limited. Some field estimates of rhyolite yield strengths based on flow morphology have been made (e.g., Hulme, 1974; Moore and others, 1978; McBirney and Murase, 1984), but comparable studies of tensile strength are unavailable. Shaw (1980) presents a curve defining a range of tensile strength values for basaltic magma as a function of crystal content. Strength increases from 10^{-4} to 10^2 bar, as crystal content goes from 20% to 100%. It is not clear how to extrapolate these data to more silicic lavas with their greater degree of polymerization. Tabulated laboratory data on *compressive* strengths of lavas (Carmichael, 1984) indicate that for a given set of pressure and temperature conditions, strength increases with silica content. If we assume that tensile strengths behave in a similar fashion, then we may consider the basaltic tensile strength values cited by Shaw (1980) as lower limits for rhyolite lavas.

In order for shear fractures to develop in a lava flow, it must

FLOW DIRECTION ⟶

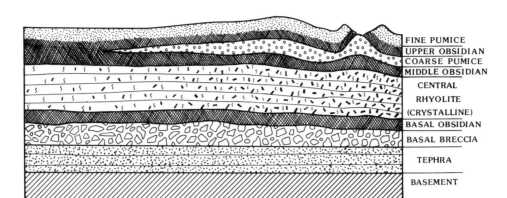

Figure 7. Schematic diagram showing formation of the coarsely vesicular pumice layer. Increasing downstream crystallization accompanied by the release of gases and formation of microcracks leads to rise of the gases to the base of the brittle surface crust, where the resultant bubbles can expand and coalesce. Thickness of this layer increases downstream, and eventually allows the rise of diapirs to the flow surface.

be capable of deforming. In any moving fluid, shear stress due to weight increases in proportion to the depth and the underlying slope (Bird and others, 1960). Thus, the thickness of the nondeforming surface crust will be defined by the depth below which the shear stress exceeds the yield strength (Fig. 8). If we assume that a rhyolite flow has a constant yield strength of 2×10^6 Nt/m^2 (Hulme, 1974; McBirney and Murase, 1984), a density of 2200 kg/m^3, and an underlying slope of 5°, the strength would only be exceeded at depths greater than 1 km. This result implies that no rhyolite flows should be able to deform at all. However, if we take into account the strong temperature dependence of yield strength (McBirney and Murase, 1984), we obtain results more consistent with observations.

Rheological measurements by Murase and others (1985) on samples of dacite from the Mount St. Helens dome show that yield strengths drop from 2.2×10^5 Nt/m^2 to zero as temperature increases from 800 to 1000°C. The inward increase in temperature in a rhyolite flow would thus result in a matching of yield strength and shear stress at depths on the order of 10 m. Lower temperatures near the base of a flow might result in a basal layer whose high yield strength also prevented deformation. In the deformable portion of the flow between the two rigid crusts, the shear stresses may also exceed the lava's tensile strength, resulting in the development of microcracks throughout much of the flow thickness.

Crystallization should begin as soon as a lava flow cools below its liquidus. However, crystallization of silicic lava is retarded by its high viscosity. The stratigraphy observed in the Inyo drill cores suggests that only in the hottest central portion of the flow was the viscosity low enough to allow sufficient ionic diffusion for crystals to grow. The shear stresses associated with flow advance drew the crystal-rich portions of the interior into flow layers. The presence of undeformed lithophysae in the glassy portion of the flow surrounding the foliated interior indicates that they formed more slowly after the flow stopped moving. Apparently, the higher viscosity of these cooler parts of the flow resulted in significantly slower diffusion and crystallization rates. The result is a foliated, crystalline, central region overlain and underlain by zones of glassy obsidian whose lithophysa content decreases upward and downward, respectively.

Crystallization of feldspar and quartz in the flow interior will release gases that can then migrate into nearby microcracks. Measurements of samples from the interior of Obsidian Dome indicate 0.2 wt % water is present in the rhyolitic glass (Westrich and others, 1985). Thus, for every cubic meter of crystallizing melt, the equivalent of approximately 4400 cm^3 of liquid water could ideally be released as vapor (assuming that biotite, hornblende, and other hydrous minerals do not form).

In basaltic lava flows, gases commonly migrate as rising,

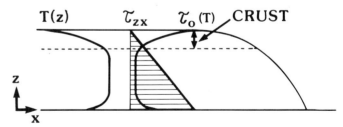

Figure 8. Schematic cross section through an active rhyolite lava flow. $T(z)$ is the temperature profile based on a simple cooling model (Fink and others, 1983) with heat loss from the upper surface by radiation and convection. T_{zx} is the shear stress, which increases downward with loading. $T_o(T)$ is the temperature dependent yield strength of the lava. Intersection of T_{zx} and $T_o(T)$ defines the base of the brittle surface crust of the flow.

coalescing bubbles (e.g., see Swanson, 1973; Peck, 1978; Sahagian, 1985), but in rhyolite flows this mechanism results in negligible movement due to the much higher viscosity of silicic lava (Sparks, 1978). However, the presence of transient fractures and microcracks allows water vapor to rise as high as the base of a rhyolite flow's upper crust. There it may pass through a series of interconnected bubbles, causing them to alternately expand and contract, eventually producing the distorted vesicular texture and breccia-filled vesicles seen in the CVP layer.

Advance and crystallization of the flow will continue to liberate more gases that will contribute to the thickening of the CVP layer. Upward migration of the vapor-equivalent of the 4400 cm^3 of water released by crystallization of a cubic metre of lava could increase the thickness of the CVP layer by from 30 cm to slightly over 1 m, depending upon the concentration and expansion of the gases. Growth of the CVP layer would be increased to a greater extent if other volatile species were also mobilized by crystallization and migrated with the water vapor. Due to the low solubility of water in rhyolite glass at elevated temperatures (Sparks, 1978), significant water resorption into the thin glass of the bubble walls will proceed only slowly during late-stage cooling of the flow. During the active lifetime of the flow, water will remain largely as free vapor, preventing collapse of the bubbles.

While gases rising from the flow interior form the CVP layer, vesiculation of the flow surface forms the FVP layer. Surface vesiculation of the flow will extend down to a depth below which the lithostatic load prevents additional exsolution of volatiles. Cooling of the upper surface also inhibits crystallization and prevents bubbles from growing very large. The product of these two effects is an aphyric FVP surface crust with vesicles whose size decreases downward toward a glassy OBS layer. Continued cooling at the surface and movement of the flow may cause large fractures to propagate down through this brittle crust to the depth at which the flow remains ductile.

This surface crust will be denser than the underlying CVP zone, due to the latter's higher water content and vesicularity. Consequently, regularly spaced diapirs of pumice may develop below the crust (Fink, 1980). Buoyant stresses associated with these diapirs could exert pressure on the base of the crust, eventually allowing the less dense material to break through to the flow surface. The rise of this pumice to the surface will allow additional expansion of bubbles and continued lowering of density, accentuating the coarsely vesicular texture. Continued addition of water vapor to the CVP layer may also lead to a significant increase in gas pressure which may result in the formation of explosion craters like those observed on the surfaces of some obsidian flows. Eichelberger (1974) concluded that because these craters never extend more than one-third of the way through a flow, and because they are often found on the uppermost of overlapping flow lobes, they must be caused by degassing of the flow interiors rather than by the heating of external water sources such as snow or ponds. He also noted that their nearly circular shape indicates that they formed relatively late during the emplacement of the lava. The absence of fragments of crystalline material from the interiors of flows around these craters indicates that they are generated no lower in the stratigraphy than the base of the CVP layer. In addition to forming craters, the CVP layer may also serve as a source for pyroclastic flows. For example, the rapid release of pressure associated with slumping of the front of a Guatemalan dacite lava flow generated several destructive pyroclastic flows in 1973 (Rose and others, 1977). Infiltration of rainwater or snowmelt into the hot interiors of the flows may have triggered the formation of the explosion craters; however, no meteoric waters seems to have been involved with the initiation of the Guatemalan pyroclastic flows (Rose and others, 1977).

The amount of CVP exposed on a flow will depend on the ability of the interior layer to rise diapirically to the surface. The buoyancy of the CVP layer will in turn depend on its thickness and density. The more crystallization that has taken place, the more released gases that will be available to increase the buoyancy of the CVP layer. Hence, for two flows of the same size but differing volatile contents, the more water-rich flow should have more CVP exposed on its surface. Conversely, flows that differ in size but not in volatile content should show a positive correlation between volume (or surface area) and amount of exposed CVP. Furthermore, flows of lower silica content that crystallize hydrous minerals should exhibit less CVP.

SUMMARY AND DISCUSSION

The model presented above qualitatively describes conditions and processes that may form the various textures observed in the drill cores through Inyo Obsidian Dome and on the surfaces of many other silicic lava flows and domes. The principal processes involved are vesiculation, crystallization, fracturing, and the migration of exsolved gases. Primary vesiculation occurs near the flow's upper surface and down to a depth controlled by lithostatic pressure. Crystallization in the flow interior is controlled by the temperature-dependent viscosity and ionic diffusion. The finite yield strength of the cooling lava results in a rigid surface crust that is carried passively on top of the moving flow. Shear stresses caused by the weight and movement of the lava result in the development of microcracks throughout the portion of the flow underlying the rigid crust. Gases released by crystallization of the flow interior migrate upward through these cracks to the base of the nondeforming crust. The anomalously high gas content of this zone below the rigid crust allows it to retain open bubbles at depths where vesiculation would normally be suppressed. Continued crystallization leads to thickening of the gas-rich layer as well as to lowering of its density, in some cases allowing it to rise buoyantly to the flow surface. Increased vapor pressure associated with the gas-rich zone may result in explosion craters on the flow surface or pyroclastic flows generated at sites of flow front collapse.

This model has implications for the assessment of hazards related to the emplacement of silicic lava flows and domes. The well-documented sequence of explosive activity followed by effusive activity results from a progressive decrease in the volatile content of the erupting magma. Hazards are greatest during the early pyroclastic activity and are generally assumed to decrease with time, especially during the quiescent emplacement of silicic domes and flows when the principal hazards are related to rockfalls at collapsing flow fronts. However, our model suggests that crystallization during flow advance can generate a gas-rich interior zone capable of producing explosive activity both on a flow surface and in a flow front. In contrast to those hazards related to the progressive decrease in magmatic water content, these hazards *increase* with duration and extent of silicic lava flows. As an example, the El Brujo dacite flow was 1.5 km long and 2 years old before collapse of its front generated pyroclastic flows (Rose and others, 1977).

Studies of the petrologic evolution of recent silicic magma systems (e.g., Bailey and others, 1976; Hildreth, 1981, 1983; Bacon and others, 1981; Mahood, 1980) require careful documentation of the emplacement sequences and volumes of young extrusions. For some systems this is complicated by difficulties in determining the boundaries of individual flows and domes. Recognition that coarsely vesicular pumice outcrops occur preferentially around the outer margins of silicic extrusions allows these boundaries to be delineated. For example, mapping the distribution of CVP outcrops on one of the Crater Glass rhyolite flows (see Fig. 2a) on the Medicine Lake Highland Volcano indicated that it was comprised of three coalesced lobes, which in turn helped define the geometry of the underlying conduit (Fink and Pollard, 1983). The continuity of long CVP ridges on the surfaces of other flows (e.g., Fig. 1) allows the determination of the sequence of emplacement of different lobes.

Our study suggests that there are advantages to studying both small and large domes. Small domes display little textural variation, commonly contain structural evidence of their vent geometry, and most likely formed during a single extrusive event. Hence they are best suited to studies of flow mechanics, the influence of conduit geometry on dome morphology, and the geochemical variation among many domes extruded from a single dike. Larger extrusions show greater textural diversity, have surfaces disrupted by a variety of flow processes and textures, and are more likely to have sampled magma whose composition was changing. Thus they contain more information about the processes that accompany surface flow of lava, the relationships between volatiles and flow textures, and the evolution of magma bodies that feed individual vents.

It has been documented that dome lavas have lower volatile contents than do the tephra whose eruption precedes them (Eichelberger and Westrich, 1981), and that the rocks with CVP texture have the highest volatile contents within the Inyo Obsidian Dome drill cores (Westrich and Eichelberger, 1984). However, the relative volatile contents of different domes emplaced from a single dike may reflect the factors that control where and when eruptions actually occur. The relationship shown in this paper between total dome volume and amount of coarsely vesicular pumice suggests that volatile content may influence the durations and volume flow rates of eruptions that produce silicic extrusions. Future studies designed to compare the actual volatile contents of such sets of extrusions with the textures they exhibit, their volumes, and their relative ages may allow determination of the specific factors that control these eruptions.

ACKNOWLEDGMENTS

We thank J. R. Holloway for valuable insights and discussions. The paper greatly benefited from reviews by J. D. Clemens, P. T. Delaney, M. C. Malin, H. R. Shaw, and an anonymous referee. Supported by Department of Energy Grant DE-FG02-85ER13320 and National Science Foundation Grant EAR-8309500.

REFERENCES CITED

Anderson, C. A., 1941, Volcanoes of the Medicine Lake Highland, California: Berkeley, University of California Publications, Bulletin of the Department of Geological Sciences, v. 25, p. 347–422.

Bacon, C. R., Macdonald, R., Smith, R. L., and Baedecker, P. A., 1981, Pleistocene high-silica rhyolites of the Coso Volcanic Field, Inyo County, California: Journal of Geophysical Research, v. 86, p. 10223–10241.

Bailey, R. A., Dalrymple, G. B., and Lanphere, M. A., 1976, Volcanism, structure, and geochronology of Long Valley Caldera, Mono County, California: Journal of Geophysical Research, v. 81, p. 725–744.

Bird, R. B., Stewart, W. E., and Lightfoot, E. N., 1960, Transport phenomena: New York, Wiley, 780 p.

Blake, S., 1981, Eruptions from zoned magma chambers: Geological Society of London Journal, v. 138, p. 281–287.

Bonnichsen, B., 1982, Rhyolite lava flows in the Bruneau–Jarbidge eruptive center, southwestern Idaho, *in* Bonnichsen, B., and Breckenridge, R. M., eds., Cenozoic geology of Idaho: Idaho Bureau of Mines and Geology Bulletin 26, p. 283–320.

Carmichael, R. S., editor, 1984, CRC handbook of physical properties of rocks: Volume III; Boca Raton, Florida CRC Press, 340 p.

Eichelberger, J. C., 1974, Origin of andesite and dacite: The petrographic and chemical evidence for volcanic contamination [Ph.D. thesis]: Palo Alto, California, Stanford University, 70 p.

Eichelberger, J. C., and Westrich, H. R., 1981, Magmatic volatiles in explosive rhyolitic eruptions: Geophysical Research Letters, v. 8, p. 757–760.

Eichelberger, J. C., Lysne, P. C., and Younker, L. W., 1984, Continental scientific drilling at Inyo Domes, Long Valley Caldera, CA [abs.]: EOS (American Geophysical Union Transactions), v. 65, p. 1096.

Fink, J. H., 1980, Gravity instability in the Holocene Big and Little Glass Mountain rhyolitic obsidian flows, northern California: Tectonophysics, v. 66, p. 147–166.

——, 1983, Structure and emplacement of a rhyolitic obsidian flow: Little Glass Mountain, Medicine Lake Highland, northern California: Geological Society of America Bulletin, v. 94, p. 362–380.

Fink, J. H., and Pollard, D. D., 1983, Structural evidence for dikes beneath silicic domes, Medicine Lake Highland, California: Geology, v. 11, p. 458–461.

Fink, J. H., Park, S. O., and Greeley, R., 1983, Cooling and deformation of sulfur flows: Icarus, v. 56, p. 38–50.

Fink, J., Manley, C., and Krinsley, D., 1985, Volatile migration, crystallization, and vesiculation during emplacement of Inyo Obsidian Dome [abs.]: EOS (American Geophysical Union Transactions), v. 66, p. 387.

Goff, F., Rowley, J., Gardner, J., Hawkins, W., Goff, S., Charles, R., Pisto, L., White, A., Eichelberger, J., and Younker, L., 1984, Valles Caldera #1, a 856-m corehole in the southwestern ring-fracture zone of Valles Caldera, New Mexico [abs.]: EOS (American Geophysical Union Transactions), v. 65, p. 1096.

Heiken, G., 1978, Plinian-type eruptions in the Medicine Lake Highland, California, and the nature of the underlying magma: Journal of Volcanology and Geothermal Research, v. 4, p. 375–402.

Hildreth, W., 1981, Gradients in silicic magma chambers: Implications for lithospheric magmatism: Journal of Geophysical Research, v. 86, p. 10153–10192.

—— , 1983, The compositionally zoned eruption of 1912 in the Valley of Ten Thousand Smokes, Katmai National Park, Alaska: Journal of Volcanology and Geothermal Research, v. 18, p. 1–56.

Hulme, G., 1974, The interpretation of lava flow morphology: Royal Astronomical Society Geophysical Journal, v. 39, p. 361–383.

Johnston, D. A., and Donnelly-Nolan, J., editors, 1981, Guides to some volcanic terranes in Washington, Idaho, Oregon, and northern California: U.S. Geological Survey Circular 838, 189 p.

Kennedy, G. C., 1955, Some aspects of the role of water in rock melts, in Poldervaart, A., ed., Crust of the Earth: Geological Society of America Special Paper 62, p. 489–504.

Mahood, G. A., 1980, Geological evolution of a Pleistocene rhyolitic center—Sierra La Primavera, Jalisco, Mexico: Journal of Volcanology and Geothermal Research, v. 8, p. 199–230.

McBirney, A. R., and Murase, T., 1984, Rheological properties of magmas: Annual Reviews of Earth and Planetary Science, v. 12, p. 337–357.

Miller, C. D., 1985, Holocene eruptions at the Inyo volcanic chain, California—Implications for possible eruptions in the Long Valley caldera: Geology, v. 13, p. 14–17.

Moore, H. J., Arthur, D.W.G., and Schaber, G. G., 1978, Yield strengths of flows on the Earth, Mars, and Moon, in Proceedings, Lunar and Planetary Science Conference, 9th: New York, Pergamon, p. 3351–3378.

Murase, T., and McBirney, A. R., 1973, Properties of some common igneous rocks and their melts at high temperatures: Geological Society of America Bulletin, v. 84, p. 3563–3592.

Murase, T., McBirney, A. R., and Melson, W. G., 1985, Viscosity of the dome of Mount St. Helens: Journal of Volcanology and Geothermal Research, v. 24, p. 193–204.

O'Neil, J. R., and Taylor, B. E., 1985, Degassing of Obsidian Dome magma: Hydrogen and oxygen isotope studies in the Inyo Dome Chain, Long Valley area, California [abs.]: EOS (American Geophysical Union Transactions), v. 66, p. 387.

Peck, D. L., 1978, Cooling and vesiculation of Alae Lava Lake, Hawaii: U.S. Geological Survey Professional Paper 935-B, 59 p.

Rose, W. I., Pearson, T., and Bonis, S., 1977, Nuee ardente eruption from the foot of a dacite lava flow, Santiaguito Volcano, Guatemala: Bulletin Volcanologique, v. 40, p. 53–70.

Sahagian, D., 1985, Bubble migration and coalescence during the solidification of basaltic lava flows: Journal of Geology, v. 93, p. 205–211.

Shaw, H. R., 1965, Comments on viscosity, crystal settling, and convection in granitic magmas: American Journal of Science, v. 263, p. 120–152.

—— , 1980, The fracture mechanisms of magma transport from the mantle to the surface, in Hargraves, R. B., ed., Physics of magmatic processes: Princeton, New Jersey, Princeton University Press, p. 201–264.

Smith, R. L., 1979, Ash-flow magmatism, in Chapin, C. E., and Elston, W. E., eds., Ash-flow tuffs: Geological Society of America Special Paper 180, p. 5–27.

Sparks, R.S.J., 1978, The dynamics of bubble formation and growth in magmas: A review and analysis: Journal of Volcanology and Geothermal Research, v. 3, p. 1–37.

Sparks, R.S.J., Huppert, H. E., and Turner, J. S., 1984, The fluid dynamics of evolving magma chambers: Royal Society of London Philosophical Transactions, ser. A, v. 310, p. 511–534.

Spera, F. J., Yuen, D. A., Kirschvink, S. J., 1982, Thermal boundary layer convection in silicic magma chambers: Effects of temperature-dependent rheology and implications for thermogravitational chemical fractionation: Journal of Geophysical Research, v. 87, p. 8755–8767.

Swanson, D., 1973, Pahoehoe flows from the 1969–1971 Mauna Ulu eruption, Kilauea Volcano, Hawaii: Geological Society of America Bulletin, v. 83, p. 615–626.

Taylor, B. E., Eichelberger, J. C., and Westrich, H. R., 1983, Hydrogen isotopic evidence of rhyolitic magma degassing during shallow intrusion and eruption: Nature, v. 306, p. 541–545.

Westrich, H. R., and Eichelberger, J. C., 1984, Water content and lithology of the Obsidian Dome flow [abs.]: EOS (American Geophysical Union Transactions), v. 65, p. 1127.

Westrich, H. R., Stockman, H. W., and Taylor, B. E., 1985, Volatile content of Obsidian Dome and the Inyo Dike [abs.]: EOS (American Geophysical Union Transactions), v. 66, p. 387.

Williams, H., 1944, Volcanoes of the Three Sisters region, Oregon Cascades: Berkeley, University of California Publications, Bulletin of the Department of Geological Sciences, v. 27, p. 37–84.

MANUSCRIPT ACCEPTED BY THE SOCIETY MAY 5, 1986

Textural heterogeneities and vent area structures in the 600-year-old lavas of the Inyo volcanic chain, eastern California

Daniel E. Sampson*
*Earth Sciences Board,
University of California,
Santa Cruz, California 95064*

ABSTRACT

The most recent eruption in the Inyo volcanic chain (600 y.B.P.) produced two chemically and mineralogically distinct lava types. The two are here distinguished by their textural characteristics and are termed coarsely porphyritic (CP) and finely porphyritic (FP). The vesicular CP lava is texturally relatively homogeneous, whereas the FP lava is markedly variable in texture. The two lava types mixed during or shortly before eruption, resulting in the formation of a small volume of hybrid lava and dramatic mixing textures. The FP magma was extruded as air-fall pumice and lava from all three vents while the CP magma primarily extruded as lava during the later stages of eruption at the southern two of the three 600-B.P. magmatic vents.

Previous workers have interpreted the 600-B.P. eruption as the result of intrusion of a rhyolitic dike along north-trending fractures. The eruptive sequence of the two lava types, combined with the along-chain and within-flow distribution of textural and chemical variations within the FP lava, indicate the following details regarding stratifications within the conduits and preeruptive dike: (1) the FP magma overlaid the CP magma with an intervening mechanically mixed zone; (2) the FP magma was itself chemically zoned both vertically and laterally, but SiO_2 generally decreased southward; and (3) the FP magma probably wedged out to the south.

The orientation of flow foliations and extrusive fractures in the vent areas of two of the three lava flows from this eruption reveal the geometry of conduits which fed the flows. The conduits for the last-erupted lava on the Deadman Creek and Obsidian flows were pipelike in shape rather than fissures, but Obsidian flow displays evidence for an originally elongate vent that probably evolved to a more pipelike shape toward the end of eruption.

INTRODUCTION

Detailed studies of the structures and textures of rhyolitic lavas are few, and most have concentrated on older, dissected flows and domes (e.g., Cole, 1970; Christiansen and Lipman, 1966). Recent work by Fink (1980a, 1980b, 1983) and Fink and Pollard (1983), however, has shown that surface mapping of uneroded Holocene rhyolite lavas can be used to decipher the eruptive history of the lavas and the geometry of conduits which fed the flows. A knowledge of conduit locations and geometries can aid in the assessment of geothermal potential, and eruptive histories can be used to assess volcanic hazards associated with potential eruptions in volcanically active regions.

The most recent volcanic event in the Long Valley area of eastern California occurred in the Inyo volcanic chain about 600 years ago (Miller, 1985). This event may serve as a useful model for predicting the nature of potential eruptions (Miller, 1985) associated with recent seismic activity and deformation in the Long Valley caldera (Fig. 1; Ryall and Ryall, 1982; Miller and others, 1982; Kerr, 1982; Cockerham and Savage, 1983). Both lava and tephra of moderate volume were erupted within the

---

*Present address: Chemistry Department, Woods Hole Oceanographic Institution, Woods Hole, Massachusetts 02543

north moat of the caldera during this 600-B.P. event (Miller, 1985). The Inyo chain has also recently been the site of shallow scientific drilling as part of the Continental Scientific Drilling Project (Eichelberger and others, 1984, 1985).

This paper presents structural and textural information gathered from surface mapping of the three lava flows produced in the 600-B.P. eruption. The textural information is then used to reconstruct the eruptive sequence of the lavas while the structural data reveals the location and geometry of vents feeding the flows. By documenting the surface lithologies and structures from this most recent Inyo eruption, this paper describes geologic relationships which can be compared and contrasted with information gathered from the subsurface through drilling. It also serves as a foundation for forthcoming petrologic work.

Previous field studies of the Inyo chain involved structural mapping of the older domes (Mayo and others, 1936), reconnaissance geology (Bailey and others, 1976; Huber and Rinehart, 1967; Bailey and Koeppen, 1977), study of the phreatic eruptions at Inyo Craters (Rinehart and Huber, 1965), tephrochronologic work (Wood, 1977), and hydration-rind dating of tephra and lava (Wood, 1983; Kilbourne and others, 1980). Miller (1985) revised Wood's (1977) tephrochronology, mapped structural features along the chain, and reported on the ages and volumes of tephra and lava for each eruption. Fink (1986) has mapped structural features along the chain and on the flows; he uses this information to deduce feeder-dike geometry. Petrologic studies in progress include Bailey (1983, 1984), Bailey and others (1983), Sampson and others (1984, 1985), and Sampson and Cameron (in prep.).

GEOLOGIC SETTING

The Inyo volcanic chain of eastern California is located at the eastern base of the Sierra Nevada escarpment along a zone of north-trending faults and ground cracks associated with east-west crustal extension (Fig. 1) (Miller, 1985; Christensen, 1966; Mayo and others, 1936). The chain consists of seven rhyolitic magmatic and numerous phreatic eruptive centers aligned in a north-trending, roughly 2-km-wide zone extending south from the south end of Mono Craters for about 11 km into the northwest moat of the Long Valley caldera (Fig. 2). Early studies do not agree on the number of magmatic vents included in the chain or on the criteria for a vent to be included (cf. Mayo and others, 1936; Jack and Carmichael, 1968). Dating of the tephras (Wood, 1977; Miller, 1985) and lavas of the region (Wood, 1986; Kilbourne and others, 1980; Bailey and others, 1976) shows that the chain comprises the rhyolites of Holocene age that lie to the south of the arcuate trend defined by the Mono Craters (Fig. 1). Subsequent studies have recognized six or fewer magmatic vents (Bailey and others, 1976; Wood, 1977, 1986; Bailey, 1980). My work has revealed the presence of a seventh rhyolite dome (informally referred to as the caldera wall dome), located about 150 m west of the southwest edge of the Glass Creek flow on the steep north-northwest wall of the Long Valley caldera, that has

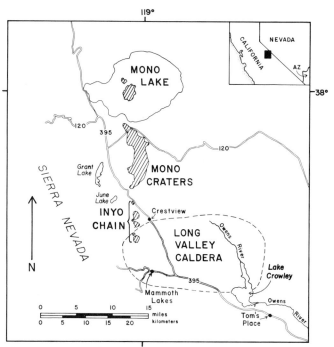

Figure 1. Index map showing location of Inyo volcanic chain, east-central California. Diagonally ruled areas are Holocene volcanics.

not been noted by previous workers (Fig. 2). It is approximately 100 by 50 m across and forms roughly half of a dome shape that protrudes to the south from the caldera wall. The caldera wall dome is the smallest in the chain and is similar in size, texture, mineralogy, chemistry, and probably age (Miller, 1985) to the small cratered dome on Glass Creek first noted by Mayo and others (1936).

Miller (1985) provided a detailed accounting of the eruptive chronology of the chain. Early eruptions date back to approximately 6000 B.P. and all produced only small lava domes with or without small volume pyroclastic deposits. The most recent eruption occurred between 550 and 650 years ago (^{14}C dates) and included pyroclastic eruptions at three vents contemporaneous with and followed closely by many phreatic explosions. This was followed by the most voluminous extrusion of lava in the history of the chain, which produced a relatively large (>1 km^2) rhyolite lava flow at each pyroclastic vent. From north to south these are Obsidian flow, Glass Creek flow, and the Deadman Creek flow (Fig. 2). These flows form about 80% of the lava volume in the chain and are the subject of this study.

The reader is referred to Bailey and others (1976) for a complete summary of the geologic and volcanic history of the region.

LAVA TEXTURES

Chemical analyses by Jack and Carmichael (1968) first illustrated that the lavas of the Inyo chain were compositionally

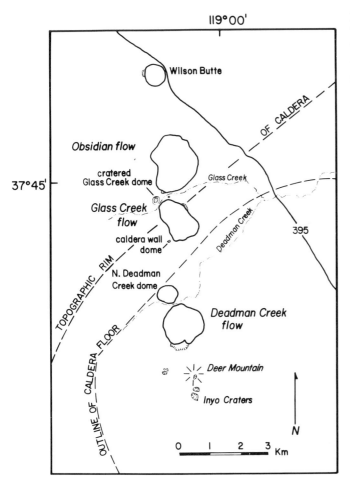

Figure 2. Map showing outlines of Holocene rhyolite domes and flows and phreatic craters (hachured) of the Inyo volcanic chain. Dashed lines indicate topographc rim and outline of the floor of the Long Valley caldera.

heterogeneous. Bailey and others (1976) subsequently clarified that two distinct rock types were intermixed in the youngest flows of the chain; one a "light, coarsely porphyritic hornblende-biotite rhyodacite," and the other a "dark, sparsely porphyritic rhyolitic obsidian."

Recent geochemical work by Sampson and others (1983) and Sampson and Cameron (1984) has shown that these two lava types are chemically distinct as well. This paper illustrates that the textural distinctions are more complex than noted above. For this study, the two lava types are called coarsely porphyritic (CP) and finely porphyritic (FP), because phenocryst size is an obvious textural distinction between them. Table 1, Figure 3, and the following descriptions provide the reader with the means to distinguish these two lava types in the field on the basis of their textural characteristics.

The CP lava is texturally and chemically nearly homogeneous, whereas the FP lava is markedly variable in appearance and chemistry (Sampson and Cameron, 1984). In the FP lava, phenocryst size and content changes slightly within a groundmass that shows a wide range in color and texture (Fig. 3a). The textural heterogeneity of the groundmass forms the basis for defining four map units within the FP lava (their distributions are shown in Fig. 4): (1) coarsely vesicular pumiceous lava (map unit cv in Fig. 4), (2) nonvesicular obsidian (map unit ob in Fig. 4), (3) finely vesicular pumiceous lava (map unit fv in Fig. 4), and (4) dense, microcrystalline lava (map unit dm in Fig. 4). Photographs of hand samples from each of these units are provided in Figure 3a. These textural variations are essentially the same as those mapped by Fink (1983) on Little Glass Mountain, a Holocene rhyolite flow in northeastern California. The following qualifications accompany the delineations of the FP textural units displayed in Figure 4: (1) the units described represent end-member textural types; textures transitional between those outlined exist, but are volumetrically minor in comparison to the four units delineated; (2) although the units on the maps are separated by solid lines, many contacts are gradational; and (3) the map areas delineate regions where a given texture is volumetrically dominant and thus not necessarily exclusive. The FP material also erupted as coarsely vesicular air-fall pumice. Though this is not displayed on the maps, the tephra ubiquitously covers the ground surface surrounding the flows. Miller (1985) provided isopach maps of the four principal tephra eruptions from the 600-B.P. event.

Vesicle shapes and sizes in the FP coarsely vesicular (FPcv) lava are notably different from those in the other vesicular FP units. Most of the vesicles in this unit are subspherical in shape, have smooth, curvilinear bubble walls, and are close in size to the mean vesicle size (Table 1). In contrast, most of the vesicles in the other map units are angular in outline, have jagged bubble walls, and show a more even distribution of vesicle size throughout the size range. The subspherical shapes, smooth bubble walls, and larger average size of the vesicles in the FPcv lava may indicate that the lava was not quenched immediately after vesiculation; this might allow bubbles to deform plastically and small bubbles to coalesce into larger ones. These observations concur with those of Fink and Manley (this volume) who interpret this textural unit to have formed within the flow as a result of upward volatile migration and entrapment near the top of the flow while much of the lava was still molten and moving. In contrast, the angular, jagged-edged bubbles common in the other units may have formed as a result of rapid inflation of a cooling, viscous material that was quenched shortly after vesiculation (such as the air-fall pumice upon eruption). In contrast to the light colors of the other vesicular units, the coarsely vesicular lava is dark gray to black. This probably reflects the lack of abundant microvesicles that are common in the lighter colored units.

Table 2 summarizes point-count data on examples of CP lava and on some of the textural units of FP lava. All of the FP textural units are sparsely porphyritic; phenocrysts (grains larger than 0.5 mm) compose about 2.5 to 3 vol% of the rock (vesicle free). Phenocryst phases, in order of decreasing abundance, include plagioclase, sanidine, biotite, Fe-Ti oxides, orthopyroxene

TABLE 1. SUMMARY AND COMPARISON OF TEXTURAL CHARACTERISTICS OF CP AND FP LAVA TYPES

	CP Lava	FP Lava			
Degree of textural variability	Slight: mostly moderately vesicular, white, abundantly and coarsely porphyritic (CP) lava	Great: obsidian (ob), finely vesicular (fv), coarsely vesicular (cv), and dense, microcrystalline (dm), finely porphyritic lava plus coarsely vesicular airfall pumice			
Chemical variation	Nearly homogenous at ~71.5 wt% SiO_2	Zoned from 70-74 wt% SiO_2			
Phenocryst mineralogy	Plag>San>Bt>Hb>Qtz>Oxides>Opx+Cpx+ Zircon+Allanite	Plag>San>Bt>Oxides>Opx or Hb>>Cpx>Zircon			
Phenocryst size	Feldspars commonly > 1 cm, mafics generally ~3 mm; Glomeroporphyries common	Mostly $\leq$2 mm; glomeroporphyries absent			
Phenocryst content (vesicle free)	25-40%	2-3%			
Map unit(s)	CP	fv	cv	ob	dm
Vesicle content	10-30%	~30%	$\geq$50%	0%	10-30%
Vesicle size range	0.1-5 mm	0.1-3 mm	0.5 mm-4 cm	n.a.	0.1-4 mm
Mean vesicle size	~1.5 mm	~0.5 mm	~0.5 mm	n.a.	~0.5 mm
Vesicle outlines	Jagged and smooth	Mostly jagged	Mostly smooth	n.a.	Mostly jagged
Groundmass color	White	Light gray to white	Dark gray to black	Black	Blue-gray
Comment	Some bands of dense black glass (see Fig. 3b)	Vesicle size evenly scattered through size range	Most vesicles close to mean size	Dense obsidian	Denser than other vesicular units

(high-silica samples) or hornblende (low-silica samples), and clinopyroxene (observed but not counted). Microphenocrysts of zircon were also observed in thin section and in heavy mineral separates. Phenocryst sizes are generally about 2 mm or less, though feldspars rarely reach 5 mm in length. FP lava samples with nearly identical mineralogies and phenocryst sizes can appear markedly different in hand sample because of the variation in groundmass texture and color and the resulting variation in contrast between phenocrysts and matrix. Feldspars, for example, tend to form the largest phenocrysts and, because they are so much more conspicuous, appear more abundant in a black obsidian than in a gray, vesicular rock, even though abundances may be nearly identical (compare I-2-3 and I-2-13 in Table 2 and in Fig. 3a). Microlite contents vary from a few to more than 15% of the rock volume. The groundmass of the air-fall pumice is nearly completely glassy, indicating that most of the groundmass crystallization occurred after the explosive phase of the eruption.

The CP lava (map unit CP in Figs. 4b and c) is texturally nearly homogeneous with only minor variations in color and vesicularity (Fig. 3b). It is more abundantly porphyritic than the FP lavas, and phenocrysts commonly are in excess of 25 vol% (excluding vesicles). Phenocryst mineralogy of the CP lava (Table 2) is similar to that of the FP lava except for the following: (1) quartz is an additional phenocryst phase, (2) hornblende is larger and more abundant than in the FP lava, (3) pyroxenes are present only in trace quantity, and (4) allanite is also a trace phase. Feldspar phenocrysts are commonly a centimeter or longer in the CP lava. Mafic minerals are usually a few millimeters long but have a larger average size (about 3 mm) than those in the FP lava (about 0.5 mm). The overall effect of these differences is to produce a rock that is both more abundantly and more coarsely porphyritic than the FP lava (cf. Figs. 3a and b).

A small amount of homogenously mixed material can be found in the 600-yr-old flows (map unit by, Fig. 4a and c). The resultant hybrid is commonly a medium-brown, moderately vesicular and coarsely porphyritic lava with a phenocryst content that is still less than that of the CP lava (Fig. 3c). Mineralogy is, of course, similar to the two lava types, but the presence of hornblende as a large phenocryst phase, combined with the distinctive brown vesicular groundmass, identifies this rock as a hybrid in the field. Based on chemical and petrographic data, hybrids of other colors and textures exist but tend to overlap in texture with the FP lavas and, as such, are more difficult to identify in the field. Chemistry (Sampson and Cameron, in prep.) and petrography reveal that this unit is composed mostly of the FP lava component.

DISTRIBUTION AND CONTACTS OF THE TWO LAVA TYPES

The CP and FP lavas are heterogeneously intermixed in the southern two of the three 600-yr-old flows, but very little homogeneously mixed (hybrid) material is present. For the most part, the two lava types remain distinct and separate and show

Figure 3. Photographs of hand specimens of different Inyo lava types; centimeter rule for scale. a: Finely porphyritic (FP) lava with maximum crystal size of about 2 mm. These four specimens represent the four FP map units in Figure 4. Clockwise from upper left they are: finely vesicular lava (map unit fv), nonvesicular obsidian (map unit ob), coarsely vesicular lava (map unit cv), and dense, microcrystalline lava (map unit dm). These specimens are from various vents, but all variations can be seen at any one vent. b: Coarsely porphyritic (CP) lava; maximum crystal length is 1 cm. These specimens are from both the Glass and Deadman Creek flows and illustrate the textural homogeneity of the lava. c: Hybrid mixture of CP and FP lavas (map unit hy) from the Deadman and Glass Creek flows; brown, vesicular, coarsely porphyritic lava; thin millimeter-sized bands of white CP lava are faintly visible.

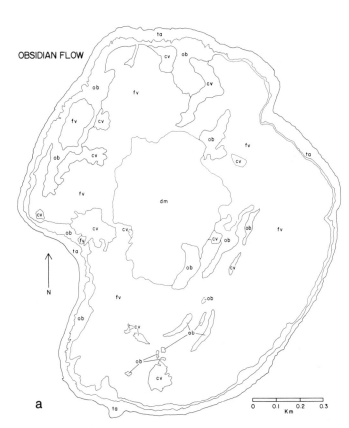
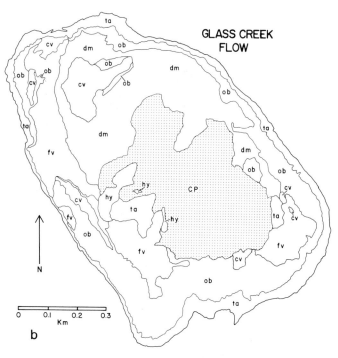
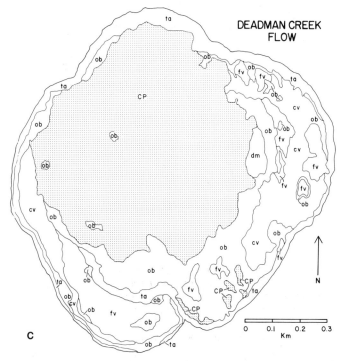

Figure 4. Maps of the distribution of the CP lava and the textural variations of FP lava for each 600-year-old flow. a = Obsidian flow; b = Glass Creek flow; c = Deadman Creek flow. CP = coarsely porphyritic (CP) lava (stipple patterned also). FP textural symbols are: ob = obsidian, fv = finely vesicular lava, cv = coarsely vesicular lava, dm = dense microcrystalline lava, ta = talus, hy = zones of intense mixing between the CP and FP lavas, where hybrid lava can be found. All FP textural units on Glass and Deadman Creek flows (except unit cv) are usually inhomogeneously mixed with blobs and bands of CP lava (see Fig. 5).

systematic variation in their distribution along the chain and within the 600-yr-old flows. The FP magma was extruded from all three magmatic vents during the 600-B.P. event. It comprises most of the juvenile tephra erupted from all three of these vents and most of Obsidian flow. On the Deadman and Glass Creek flows, the FP lava is absent in the central vent areas but abundant in the mixed flow-peripheries (Fig. 5). Both the relative abundance and absolute volume of FP lava decrease southward among the three flows (Fig. 5).

The CP lava is found in the homogeneous central areas and mixed margins of the Deadman and Glass Creek flows (Fig. 5) and as mixed blobs in the last bombs ejected from the Glass Creek flow vent (Miller, 1985). In the mixed outer portions of these two flows, the CP lava occurs as blobs and bands enclosed within and, for the most part, inhomogeneously mixed with FP lava (Fig. 6). The CP lava is absent from Obsidian flow except as a component in the miniscule volume of hydrid lava present there. It is difficult to quantitatively estimat CP vs. FP lava volumes on the two mixed flows because their relative proportions are so

TABLE 2. PETROGRAPHIC DATA ON REPRESENTATIVE INYO LAVAS

Specimen Number	Unit	Number of counts	G-mass (%)	vesicles (%)	Plag (%)	K-spar (%)	Qtz (%)	Bt (%)	Hb (%)	Opx (%)	Cpx (%)	Opaq (%)
I-2-3	FPob	2000	97.5	----	1.8	0.5	----	0.1	----	0.1	----	0.2
I-2-3	FPdm	2000	83.0	15.0	0.9	0.9	----	0.5	----	0.1	----	0.3
	vesicle free =		97.0		1.1	1.0	----	0.5	----	0.1	----	0.3
I-2-14	FPcv	3025	46.0	53.0	1.2	0.1	----	0.1	----	----	----	----
	vesicle free =		97.2		2.5	0.2	----	0.07	----	----	----	----
I-5-19	FPcv	2000	52.0	47.0	0.2	0.4	----	0.4	----	0.4	----	0.3
	vesicle free =		97.5		0.4	0.8	----	0.8	----	0.4	----	0.3
I-3-24	FPob (hy)	2000	95.3	----	2.7	0.8	0.1	0.8	0.3	0.05	----	----
I-3-2	CP	2000	63.3	15.1	12.9	4.1	1.7	1.9	0.5	----	----	0.4
	vesicle free =		74.5		15.2	4.8	1.9	2.2	0.6	----	----	0.4
I-5-17	CP	2000	57.6	12.9	14.8	9.3	0.5	2.4	1.9	----	----	0.8
	vesicle free =		66.0		17.0	10.6	0.6	2.8	2.2	----	----	0.9

variable in the mixed zones. However, the volume of nonmixed CP lava clearly increases southward, composing less than 50% of the volume of the Glass Creek flow and well above 50% of that of the Deadman Creek flow (Fig. 5).

The contact between the two lava types is usually sharp and displays mixing textures (Fig. 6). On the Deadman and Glass Creek flows, there is little outcrop of FP lava that is free of inclusions of CP lava, yet large masses of homogeneous CP lava make up the central area of these flows (Fig. 5). As a result, a line can be drawn separating homogeneous CP lava from mixed zones (Figs. 4b, c, and 5). On a traverse from homogeneous CP lavas toward the mixed zone, the first indications of mixing are bands of FP lava that apparently intrude the CP lava (Fig. 6a). The bands are thin (a few centimeters) where first encountered, continuous for many meters, and nearly vertical. A few meters farther along the traverse, they begin to connect with one another and isolate subangular to subrounded blobs of CP lava as much as 2 m across, and leave them floating in a mass of FP lava (Fig. 6b). Near the contact of the two lavas the proportion of FP lava generally increases toward the flow margin. On the outlying portions of the flows, however, the relative abundance of CP lava varies irregularly, rather than simply decreasing with distance from the vent. Isolated blobs of CP lava often appear disaggregated, possibly as a result of shear rotation during flow. This is evidenced by the presence of thin, wispy bands of CP lava that trail off either side of a blob and are deformed into swirllike patterns in the FP host (Fig. 6b). Farther along the traverse, the CP blobs become more deformed until, in places, they have been completely spread out into long, continuous bands. Here the crystals of the CP lava appear to float off into the FP matrix (Fig. 6c).

It is apparent from the mixing textures that the CP lava behaved more viscously during mixing than did the FP lava. This could be solely the result of its higher crystal content. However, it

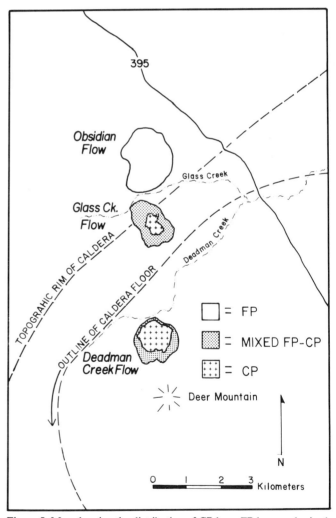

Figure 5. Map showing the distribution of CP lava, FP lava, and mixed CP-FP zones in the 600-yr-old flows in relation to the northern topographic rim of the Long Valley caldera.

Figure 6. Photo sequence of CP-FP contact and mixing zones on the Deadman Creek flow. a: Near the contact of the two units where blobs of CP (white) are isolated from one another by the intruding FP (black) into subangular to subrounded shapes; homogeneous CP lava can be seen behind and to the left of person. b: A few tens of metres from the contact of the two lava types where blobs of CP are spread out into swirls and lenses; arrow points to hammer for scale c: Close up of thin bands of disaggregating CP lava; crystals from CP are floating in the FP matrix. Figure 6b from Sampson, D. E., 1985, EOS (American Geophysical Union Transactions), v. 66, p. 712, 1985 copyright by the American Geophysical Union.

may also be due to lower temperature, greater vesicularity, or higher silica content (all of which would contribute to a higher viscosity) of the CP liquid during mixing. Major-element analysis of bulk glass separated from the CP lava shows that the CP liquid had a higher SiO_2 content (75.5 wt%) than the most silicic FP lava, which at 73.5 wt% SiO_2 is close to aphyric (and thus nearly a liquid composition). In addition to the textural evidence, chemistry (Sampson and Cameron, in prep.) illustrates that the mixing of these two magmas was more a disaggregation and incorporation of viscous CP magma by and into the more liquid FP magma than a simple two-liquid mixing. Hybrid mixtures are composed mostly of the FP lava component, and not vice versa, as if the CP magma were behaving more as an assimilant than as a liquid. In these mixed zones, obsidian is the most common texture of the FP lava, providing a marked contrast in vesicularity and color between the two units. This results in dramatic displays of the mixing textures (see Fig. 6).

The CP-FP hybrid lava (map unit hy) is more abundant on the Glass Creek flow than on the other two. On the Glass Creek flow it is most abundant near the contact of the CP-rich central area with the rest of the flow (Fig. 4b). It is restricted to small, isolated localities on the Deadman Creek and Obsidian flows. On the latter, blobs of the material can be found in the vent area of the flow as groups or isolated bodies that often have sharp contacts with the FP host. In total they make up less than 1% of the outcrop area of Obsidian flow.

DISCUSSION OF TEXTURAL DISTRIBUTION

The distributions and contact relationships of the FP textural units in the 600-year-old Inyo flows are quite similar to those found by Fink (1983) on Little Glass Mountain. This is especially true for Obsidian flow, where the CP lava is absent. Fink and Manley (this volume) review the map patterns of textural variations in rhyolite flows and show that such a variety and arrangement of textures is common.

A general eruptive scheme can be deduced for the 600-B.P. Inyo eruption using the preceding information on the distributions of the two lava types and various textures of the FP lava. The first magmatic ejecta was nonmixed, vesicular, and explosively volatile-charged FP pumice. At the southern two vents, a fluid, nonhomogenized mixture of vesicular CP and relatively degassed FP lava then flowed out over the pumiceous FP tephra deposits. The FP lava in the upper portion of this mixture continued to exsolve volatiles and formed a finely vesicular carapace on the flow (Fink, 1983). The last FP lava to erupt was denser (though slightly vesicular) and more microcrystalline than previous eruptives. In places this material was nearly homogeneously mixed with CP lava to form a brown to gray, vesicular, porphyritic hybrid rock. This was followed by the viscous effusion of a relatively homogeneous, vesicular CP lava that piled high over the vent area. The very last CP lava to erupt at the Deadman Creek vent is foliated with bands that are very coarsely and irregularly vesicular.

This eruptive sequence implies that the magmas were stratified in the conduit prior to eruption. A volatile-rich FP magma at the top of the body (represented by the air-fall pumice) graded down into a less volatile charged FP magma (the FP lavas), thence down into a mechanically mixed CP-FP magma zone. Finally, this was underlain by relatively homogeneous and vesicular CP magma.

Miller (1985), Fink and Pollard (1983), Fink (1986) and the Inyo domes drilling (Eichelberger and others, 1985) provided compelling evidence that a dike or series of dikes fed all three magmatic vents in the 600-B.P. eruption at Inyo. This evidence then allows us to also consider stratifications along the length of this dike as they are reflected in the textural and chemical variations among the three vents.

As Bailey and others (1976) have previously noted, the proportion of CP lava in these flows increases southward into the Long Valley caldera (Fig. 5). In addition, the volume of FP eruptives (lava plus juvenile tephra) decreases southward among these three vents (Miller, 1985). If the top of the dike was roughly horizontal (that is, any deviations from the horizontal were small in comparison with the vertical thickness of the magma body), and the dike was tapped from the top down by each vent, then the overlying FP magma may have thinned to the south with the boundary between the two magma types dipping to the north. Alternatively, the top of the dike may have dipped to the south (as suggested by Fink [1986] on the basis of ground-crack spacings), with the boundary between the two magmas being horizontal. In the latter case, each conduit southward could have tapped successively deeper levels of the dike than the one north of it.

The variations in the textures of the FP lavas do not correlate regularly with the variations seen in their major- and trace-element chemistries. For instance, a coarsely vesicular lava and an obsidian from one flow may be identical in chemistry, whereas two obsidians from the same flow may be very different chemically. It is also apparent from chemical evidence that the first erupted material (the air-fall pumice) is not the highest silica of

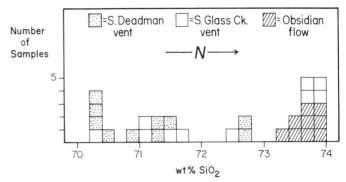

Figure 7. A histogram of silica contents in the FP lavas and tephras analyzed from the 600 y-B.P. eruption. CP and hybrid lavas are excluded. Note the general northward increase in silica content among the three vents.

the FP magma but, rather, some of it is the lowest (Sampson and Cameron, in prep.). Though most documented zoned silicic eruptions are far larger in volume than this one, most show a sequence where the opposite occurs; i.e., the first erupted material is the most silicic, and later erupted magma becomes increasingly mafic (Hildreth, 1981). There is some suggestion that the FP lava from this event becomes, on the average, more silicic northward as well (Fig. 7) (Bailey and others, 1983; Sampson and Cameron, in prep.). There are certainly exceptions to this, especially as revealed in the Inyo domes drilling (Stockman and others, 1984, and 1985), but to date none of the highest silica material has been sampled from the Deadman Creek flow, whereas the material composes nearly the entire surface of Obsidian flow. Clearly, the FP magma was not simply chemically zoned with a higher silica cap immediately prior to eruption. Neither can the complex distribution of chemistry along the chain be reconciled merely by assuming an inversely stratified body with a southward-dipping top. Rather, the geographic correlation with chemistry suggests that a significant north-south (or lateral) zonation existed in the linear magma body before it erupted, whereas the apparently cyclic variation in chemistry observed by Stockman and others (1985) and C. D. Miller (1985, personal commun.) suggests either complex stratification or that the eruptions did not tap the dike(s) in a simple top-down manner.

Because the CP and FP magmas are, for the most part, not a homogenized mixture, mixing between the two probably began shortly before eruption and may have contributed to the triggering of the eruption if a temperature contrast existed between the two magmas (causing sudden volatile exsolution in the hotter magma on contact). The CP lava is only found within the topographic confines of the Long Valley caldera, its relative proportion increases toward the interior of the caldera, and it is chemically similar to older Long Valley lava. Thus, its source may reside within the confines of the caldera (Bailey and others, 1976; Sampson and Cameron, in prep.). The northward increase in volume of the FP lava could likewise indicate that its source lies to the north of the caldera and that the two magmatic systems were thus laterally juxtaposed before intersection took place. However, further petrologic studies are necessary to determine the validity of such a model.

LAVA STRUCTURES AND VENT GEOMETRIES

Studies of the structures of lavas can lead to inferences concerning the mode of origin of the structures and the extrusive style of the lavas. Fink (1983) has used structural data to deduce flow mechanisms, lava rheology, and the styles of deformation of rhyolite flows. Fink and Pollard (1983) proposed that structural data from the vent areas of rhyolite lava flows can also reveal the geometry of an underlying conduit. They suggested that flow foliations form parallel to the walls of the conduit, and that fractures form in the vent area due to thermal contraction and the extension caused by magma pushing up from below. Therefore, a concentric arrangement of flow foliations, combined with a radial fracture pattern in the vent area lava, is the result of extrusion through a circular orifice. A predomination of linearly aligned fractures, coupled with foliations striking parallel to them, would be due to eruption through a linear vent parallel to the strike of the foliations. In this study, orientations of the flow foliations and fractures in the vent areas of the three lava flows from the 600-B.P. eruption in the Inyo chain were recorded with the aim of precisely locating vents and deciphering their geometries.

The 600-B.P. event produced a far larger volume of erupted material (approximately 0.5 km^3) than any preceding events in the chain (Miller, 1985). The lava flows covered and breached most of their tephra rings and possibly overrode numerous phreatic explosion craters (all of the flows partially overrode craters; Fig. 2). These flows were described by Mayo and others (1936) as "great, sprawling, shield shaped masses with abrupt edges." They also noted the presence of a "raised, central boss or group of spires" which "marks the position of the orifice." They called attention to a region of parallel, north-trending fissures south of and trending parallel to the chain that they thought might be genetically related to the rhyolitic eruptions. Since then, Fink and Pollard (1983), Miller (1985), and Fink (1986) have postulated that these fissures and a series of north-trending graben structures were created by intrusion of a north-trending dikelike body of magma that acted as the feeder for this 600-B.P. eruption. Additional support for this hypothesis has come through the Inyo domes drilling along Glass Creek between Obsidian and Glass Creek flows as part of the Continental Scientific Drilling Project. This drill hole intersected a rhyolitic intrusion in the postulated location of the proposed dike (Eichelberger and others, 1985). Fink (1983) and Fink and Pollard (1983) proposed that, with dikelike feeders, each lava extrusion may erupt through a fissure-shaped vent aligned parallel to the dike. They cited lava-fracture orientations and their influence on flow foliation orientations as evidence for such vent geometries at other localities (i.e., Medicine Lake, northeastern California).

Data sites reported in this study (Fig. 8) are scattered because: (1) not all of the lava was foliated, and (2) only large outcrops that appeared to be connected to the mass of the flow and that had regionally consistent foliation orientations were measured. Loose, rafted blocks were thus avoided. The attitudes were generally continuous from one outcrop to another in the center of the vent areas. Most of the outlying flow is covered by loose, blocky material and, at close range, yields little or no coherent structural information. On air photos, however, subparallel, curving ridges are apparent outside of the vent areas that may represent compressional flow folds (Figs. 8a and c) (Fink, 1980b, 1986). The topographic break between the elevated vent areas and the lower lying flow margins seems to represent a structural break as well, because structures (such as fractures or extrusive ridges) are rarely continuous across this boundary. This break probably indicates a distinct difference in the behavior of the lava during the early and later parts of the eruption, the early erupted lava behaving more fluidly, as suggested by Fink (1983).

In the center of the vent areas of Obsidian and Deadman Creek flows, the foliations strike concentrically around a point that presumably marks the center of the orifice (Figs. 8a, c, and 9). On Obsidian flow, this is in the center of the depressed area that Eichelberger and others (1985) postulated was created by magma sagging back down the vent at the close of eruption (Fig. 9). It is also precisely over the conduit intersected by the Inyo domes drilling slant hole on Obsidian flow (Eichelberger and others, 1985). Near the center of the vent area, the foliations dip shallowly away from this central point; proceeding outward from the center, the foliations progressively steepen and overturn to dip back toward their source (Figs. 8a and c). On the Deadman Creek flow, these foliations maintain a concentric strike pattern well outside of the vent area up to slightly beyond the contact between the two lava types (Figs. 4c and 8c). This foliation pattern suggests a concentrically layered (onion-skin), dome-like extrusive structure (Fig. 9), which in turn reflects extrusion of the last-erupted lava through a circular orifice rather than through a fissure (Fink and Pollard, 1983).

Such a vent geometry is also evidenced on the Deadman Creek flow by a radial fracture pattern that, like the concentric foliation pattern on this flow, extends well outside the center of the vent area (Fig. 8c). On Obsidian flow, the fracture pattern also follows the concentric foliation pattern in that it only appears radial in a small area near the center of the vent area. Both of these flows erupted onto relatively flat terrain, so their fracture patterns should reflect the geometry of their underlying conduits (Fink and Pollard, 1983). The foliations strike nearly perpendicular to these fractures, contrary to what Fink and Pollard (1983) would predict if the lava had extruded through a linear vent aligned parallel to any of the fractures (Fig. 8).

There is, however, some evidence for an elongate vent on Obsidian flow. The topographically higher flow center and the outline of the flow are elongate in a roughly north-south direction. In addition, compressional flow-folds on the east side of the flow that lie outside of the topographic break are concentric around the flow center and outline the elongate shape of the vent

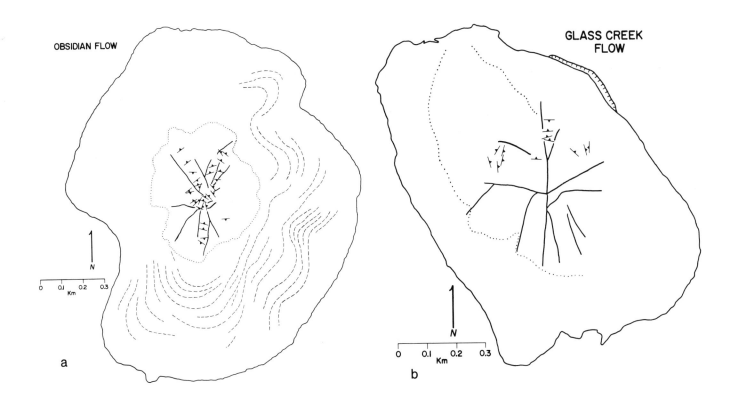

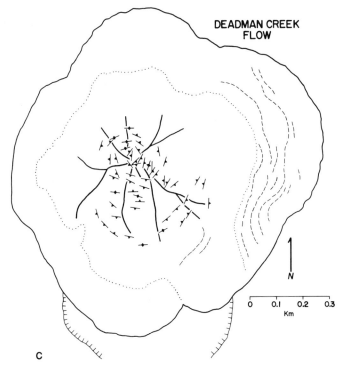

Figure 8. Maps of fractures and foliations in the vent area and compressional ridges on the outer portion of each 600-yr-old flow; a:, b:, and c: as in Figure 4. Dotted lines mark topographic breaks; elevated part is toward the flow center, solid lines are lava fractures, and foliation strike and dip symbols show the orientation of lava foliations in the extrusive ridges between fractures. Dashed lines outside the topographic break are drawn along the crest of compressional flow-fold ridges on the outer portion of the flow. Only the sense of dip is given on the foliation orientations but most are close to vertical, except near the center of the vent areas where they reach values as low as 45°.

Figure 9. Photographs taken in the center of the vent area of Obsidian flow illustrating the onion-skin structure of the last erupted lava. a: View toward south; foliation dips radially away from viewer. b: View toward north; same foliation dip pattern. Arrow in each photo points to person for scale.

area (Fig. 8a). Though flow and vent area shape are subject to topographic control and therefore do not necessarily reflect vent geometry, Obsidian flow lies on fairly flat ground. Thus it may be that early extrusion was through an elongate vent or series of aligned vents and, during eruption, the vent opened in an east-west direction and evolved to the more pipelike geometry revealed in the internal structure of the last-erupted lava.

There is circumstantial evidence for multiple vents beneath the Deadman Creek flow that align north-northwest. At the south-southeast edge of the flow where, in map view, the flow edge makes a sharp northward indentation, a large outcrop of CP lava underlies the FP-CP mixture. On the flow surface northeast of this there are scattered outcrops of nonmixed CP lava that protrude through the overlying mixed mass (Fig. 4c). This region overlying the large outcrop of CP lava appears topographically perched relative to the part of the flow lying immediately west-southwest of it (Fig. 8c; also note the talus apron within the flow south of the vent area in Fig. 4c which marks this topographic break). These relationships might be explained by a small amount of CP lava issuing from a late extrusive vent beneath the earlier emplaced FP-CP mixture, thereby perching this part of the flow above the surrounding flow.

On the Glass Creek flow, fracture patterns were probably strongly controlled by the underlying topography. This lava flowed down either side of the north rim of the Long Valley caldera and, because these fractures form to accommodate expansion (Fink, 1983; Fink and Pollard, 1983), the deepest, most prominent fractures trend east-northeast, roughly perpendicular to the principal flow directions (Fig. 8b). There is also a very prominent north-trending fracture on this flow (Fig. 8b). It is unlikely that this was formed by extrusion through a vent elongate in this direction, because foliations strike nearly normal to the trend of the fracture (Fig. 8b), rather than parallel to it, as would be predicted with such a vent geometry (Fink, 1983; Fink and Pollard, 1983). The limited foliation orientation data from this flow suggest a concentric pattern as well, though they are too sparse to be conclusive.

Thus, the surface structural evidence implies that pipelike conduits, rather than fissure-shaped vents, fed the last erupted lava on Obsidian flow and most of the lava on the Deadman Creek flow. Obsidian flow may have issued from a vent that began as a linear feature and evolved to a more pipelike shape toward the end of extrusion. The Deadman Creek flow appears to have been fed by a pipelike conduit throughout most of its extrusion, but may have been fed by two such conduits toward the end of eruption. The structural data on the Glass Creek flow are less definitive due to a lack of foliation in the vent area and due to the topographic control on fracturing.

ACKNOWLEDGMENTS

Ken Cameron and Mark Reagan provided stimulating discussions and helpful reviews throughout the evolution of the manuscript. Benjamin C. Shuraytz, Michael C. Malin, and C. Dan Miller reviewed the manuscript and provided comments and suggestions that greatly improved the paper. This work was supported by a Geological Society of America Penrose grant and by grants from the Institute for Geophysics and Planetary Physics of Los Alamos National Laboratories and the University of California.

REFERENCES CITED

Bailey, R. A., 1980, Structural and petrologic evolution of the Long Valley, Mono Craters and Mono Lake volcanic complexes, eastern California: EOS (American Geophysical Union Transactions), v. 61, p. 1149.

—— , 1983, Postcaldera evolutions of the Long Valley magma chamber, eastern California: EOS (American Geophysical Union Transactions), v. 64, p. 889.

—— , 1984, Chemical evolution and current state of the Long Valley magma chamber, in Proceedings of Workshop XIX, active tectonic and magmatic processes beneath Long Valley caldera, eastern California: Volume I: U.S. Geological Survey Open-File Report 84-939, p. 24–40.

Bailey, R. A., Dalrymple, G. B., and Lanphere, M. A., 1976, Volcanism, structure and geochronology of the Long Valley caldera, Mono County, California: Journal of Geophysical Research, v. 81, p. 725–744.

Bailey, R. A., and Koeppen, R. P., 1977, Preliminary geologic map of the Long Valley caldera, Mono County, California: U.S. Geological Survey Open-File Map number 77-468, scale 1:62,500.

Bailey, R. A., Macdonald, R. A., and Thomas, J. E., 1983, The Inyo-Mono craters: Products of an actively differentiating rhyolite magma chamber, eastern California: EOS (American Geophysical Union Transactions), v. 64, p. 336.

Christensen, M. N., 1966, Late Cenozoic crustal movements in the Sierra Nevada of California: Geological Society of America Bulletin, v. 77, p. 163–182.

Christiansen, R. L., and Lipman, P. W., 1966, Emplacement and thermal history of a rhyolite flow near Fortymile Canyon, southern Nevada: Geological Society of America Bulletin, v. 77, p. 671–684.

Cockerham, R. S., and Savage, J. C., 1983, Earthquake swarm in Long Valley, California January, 1983: EOS (American Geophysical Union Transactions), v. 64, p. 890.

Cole, J. W., 1970, Structure and eruptive history of the Tarawera volcanic complex: New Zealand Journal of Geology and Geophysics, v. 13, p. 879–902.

Eichelberger, J. C., Lysne, P. C., and Younker, L. W., 1984, Research drilling at Inyo domes, Long Valley caldera, California: EOS (American Geophysical Union Transactions), v. 65, p. 721–724.

Eichelberger, J. C., Lysne, P. C., Miller, C. D., and Younker, L. W., 1985, Research drilling at Inyo domes, California: 1984 results: EOS (American Geophysical Union Transactions), v. 66, p. 186–187.

Fink, J. H., 1980a, Gravity instability in the Holocene Big and Little Glass Mountain rhyolitic obsidian flows, northern California: Tectonophysics, v. 66, p. 147–166.

—— , 1980b, Surface folding and viscosity of rhyolite flows: Geology, v. 8, p. 250–254.

—— 1983, Structure and emplacement of a rhyolitic obsidian flow: Little Glass Mountain, Medicine Lake Highland, northern California: Geological Society of America Bulletin, v. 94, p. 362–380.

—— , 1985, The geometry of silicic dikes beneath the Inyo domes, California: Journal of Geophysical Research, v. 90, p. 11,127.

Fink, J. H., and Pollard, D. D., 1983, Structural evidence for dikes beneath silicic domes, Medicine Lake Highland, California: Geology, v. 11, p. 458–461.

Hildreth, W., 1981, Gradients in silicic magma chambers: Implications for lithospheric magmatism: Journal of Geophysical Research, v. 86, p. 10153–10192.

Huber, N. K., and Rinehart, C. D., 1967, Cenozoic volcanic rocks of the Devil's Postpile quadrangle, eastern Sierra Nevada, California: U.S. Geological Survey Professional Paper 554-D, p. D1–D21.

Jack, R. N., and Carmichael, I.S.E., 1968, The chemical "fingerprinting" of acid volcanic rocks: California Division of Mines and Geology Special Report 100, p. 17–32.

Kerr, R. A., 1982, Volcanic hazard alert issued for California: Science, v. 216, p. 1302–1303.

Kilbourne, R. T., Wood, S. H., and Chesterman, C. W., 1980, Recent volcanism in the Mono Basin–Long Valley region of Mono county, California, in Sherburne, R. W., Mammoth Lakes, California earthquakes of May 1980: California Division of Mines and Geology Special Report 150, p. 7–22.

Mayo, E. B., Conant, L. C., and Chelikowsky, J. R., 1936, Southern extension of the Mono Craters, California: American Journal of Science, v. 32, p. 81–91.

Miller, C. D., 1985, Chronology of Holocene eruptions at the Inyo volcanic chain, California—Implications for possible eruptions in Long Valley caldera: Geology, v. 13, p. 14–17.

Miller, C. D., Mullineaux, D. R., Crandell, D. R., and Bailey, R. A., 1982, Potential hazards from future volcanic eruptions in the Long Valley–Mono Lake area, east-central California and southwest Nevada—A preliminary assessment: U.S. Geological Survey Circular 877, 10 p.

Rinehart, C. D., and Huber, N. K., 1965, The Inyo Crater lakes—A blast in the past: California Division of Mines and Geology Mineral Information Service, v. 18, p. 169–172.

Ryall, F., and Ryall, A., 1982, Propagation of effects and seismicity associated with magma in the Long Valley caldera, eastern California: Earthquake Notes, v. 53, p. 46–47.

Sampson, D. E., Ardito, C. P., Kelleher, P. C., and Cameron, K. L., 1983, The geochemistry of Quaternary lavas from the Inyo-Mono Chain: Evidence for several magma types: EOS (American Geophysical Union Transactions), v. 64, p. 889.

Sampson, D. E., and Cameron, K. L., 1986, The geochemistry of the Inyo volcanic chain: Multiple magma systems in the Long Valley region, eastern California: Journal of Geophysical Research (in review).

Sampson, D. E., Williams, R. W., and Gill, J. B., 1985, ^{238}U series disequilibria in the Inyo drill core and surface samples: EOS (American Geophysical Union Transactions), v. 66, p. 388.

Sampson, D. E., Williams, R. W., Robin, M., and Gill, J. B., 1984, The age of rhyolite magma at eruption: Inyo volcanic chain, eastern California: EOS (American Geophysical Union Transactions), v. 65, p. 1128.

Stockman, H. W., Westrich, H. R., and Eichelberger, J. C., 1984, Variations in volatile and non-volatile components in Obsidian dome: EOS (American Geophysical Union Transactions), v. 65, p. 1127.

Stockman, H. W., Westrich, H. R., and Miller, C. D., 1985, Geochemistry of Obsidian dome and the Inyo dike: An overview: EOS (American Geophysical Union Transactions), v. 66, p. 385.

Wood, S. H., 1977, Distribution, correlation, and radiocarbon dating of late Holocene tephra, Mono and Inyo craters, eastern California: Geological Society of America Bulletin, v. 88, p. 89–95.

—— , 1983, Chronology of late Pleistocene and Holocene volcanics, Long Valley and Mono basin geothermal areas, eastern California: U.S. Geological Survey Open-File Report 83-0747, 84 p.

MANUSCRIPT ACCEPTED BY THE SOCIETY MAY 5, 1986

Printed in U.S.A.

Types of mineralization related to fluorine-rich silicic lava flows and domes

Donald M. Burt
Michael F. Sheridan
Department of Geology
Arizona State University
Tempe, Arizona 85287

ABSTRACT

Several types of mineralization appear to be related to the emplacement of fluorine-rich silicic lava flows and domes. An important example is the beryllium deposit at Spor Mountain, west-central Utah, where bertrandite, fluorite, amorphous silica, and Mn-Fe oxides replace dolomite fragments in tuffaceous surge deposits just beneath a topaz-bearing rhyolitic lava flow. The Be-mineralized zone is also highly enriched in F, Sn, W, Nb (and presumably Ta), Zn, Pb, and several other metals (but not in Mo). The uniform lateral character of the mineralization, the restriction of Be mineralization to the uppermost few meters of tuff, and the lack of mineralization in fluorite-bearing breccia pipes (tuffaceous vent breccias, in some cases) in underlying dolomite suggest (Bikun, 1980) that the beryllium mineralization resulted from the devitrification of the overlying lavas (a "steam iron" model). Mass-balance calculations based on comparisons of the chemical compositions of glassy and devitrified rhyolite are consistent with this unconventional interpretation.

A second important example is provided by "Mexican-type" fumarolic tin deposits, characterized by cassiterite in carapace breccias of rhyolitic domes. Low-temperature dissolution and reworking of early fumarolic cassiterite may produce the colloform "wood tin" common in this deposit type. Deposits of this type occur in Nevada and New Mexico, as well as in many areas of northern Mexico.

Fluorine-rich intrusive domes (better known as plutons) may also host metal mineralization, generally of the porphyry type. Examples include the well-known Climax-type porphyry molybdenum deposits of Colorado, New Mexico, and Utah, and the porphyry tungsten deposit at Mount Pleasant, New Brunswick, Canada. These subvolcanic deposits, in common with those associated with extrusive silicic lava flows and domes, are believed to have been derived by the crystallization-devolatization of highly fractionated magma.

INTRODUCTION

For the past several years we have been studying types of mineralization related to the emplacement of fluorine-rich silicic lava flows and domes. The fluorine-rich character of these lavas is indicated by the presence of topaz ($Al_2SiO_4F_2$) in vapor-phase cavities and along fractures. The presence of topaz leads to the term topaz rhyolite for these rocks. An alternative term is "rare metal rhyolites," in recognition of the extreme enrichment in lithophile ("fluorophile") rare metals (Li, Rb, Cs, Nb, Ta, Sn, W, U, etc.) that accompany the fluorine. These enrichments are believed to be evidence of extreme fractional crystallization of rhyolitic magma, initially somewhat enriched in these elements due to partial melting of a Precambrian crustal source in an environment of high heat flow and extension (Burt and others, 1982).

Fluorine-rich rhyolites are interesting in a volcanologic context because fluorine is believed to markedly lower the viscosity and minimum melting temperature of rhyolitic magma (Wyllie, 1979; Manning, 1981; Dingwell, 1985; Dingwell and others, 1985). In this regard, the effect of fluorine is similar to that of water; the difference is that fluorine-rich, water-poor magmas are much less likely to erupt explosively and produce pyroclastic deposits than are their water-rich counterparts, unless they interact with external sources of water (as discussed below).

Much of the work reported herein was published (Burt and Sheridan, 1981; Burt and others, 1982, Christiansen and others, 1983) in journals not closely related to volcanology; these publications did not emphasize the importance of volcanological features in localizing mineralization. In addition, field studies related to this work were all restricted to the western United States. During 1983 and 1984 we made several visits to topaz rhyolites in Mexico, mainly in and near the states of San Luis Potosi and Guanajuato. The purpose of this review paper is to summarize our previous work in the United States and to report our new observations in Mexico, with emphasis on volcanological features.

SPOR MOUNTAIN, UTAH

In economic terms, the most important example of mineralization related to a topaz rhyolite is probably the beryllium deposit at Spor Mountain, west-central Utah (Staatz and Carr, 1964; Lindsey, 1977, 1979, 1982; Bikun, 1980; Christiansen and others, 1984). Two very similar but distinct dome complexes of topaz rhyolite lava occur near the beryllium mine (Fig. 1). The older, to which mineralization is related, is the Spor Mountain rhyolite, approximately 21 Ma. The original thickness and extent of this lava is unknown; it presently is preserved in a number of small eroded fault blocks, mainly at the western foot of Spor Mountain. The lava remnants are partly buried by alluvium and valley fill. The rhyolite lava and underlying tuff rest on a sequence of lower Paleozoic carbonate rocks that make up the bulk of Spor Mountain. An apparent vent area is exposed in the farthest north of the mine workings, the Taurus pit. Higher on Spor Mountain, where the volcanic rocks have been eroded, several fluorite mines occur in breccia pipes that pass through dolomite; these breccia contain altered volcanic ash in the matrix and may represent additional vent areas.

To the east of Spor Mountain, a thick sequence of flow-banded topaz rhyolite lava flows make up the Thomas Range. These flows apparently erupted from multiple vents and varied widely in volume, as recognized from mapping of foliation and patterns of pressure ridges (Lindsey, 1982, p. 26–28). This formation, famous among mineral collectors for more than 100 y, is named for Topaz Mountain, a prominent peak at the southern end of the Thomas Range. It was earlier believed to correlate with the topaz rhyolite of Spor Mountain, but radiometric dating revealed that its age is only 6 to 7 Ma (Lindsey, 1977, 1979). This age difference, in combination with other factors discussed below, helps explain the lack of economic mineralization associated with

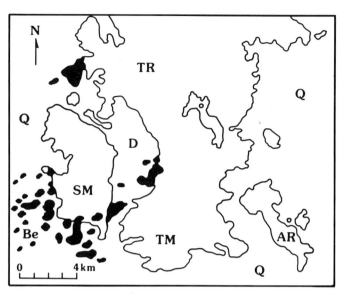

Figure 1. Generalized map of the Spor Mountain area, Utah (after Lindsey, 1979, 1982). TR = Thomas Range, the main mountain range, mainly made up of 6 to 7 Ma Topaz Mountain rhyolite (TM = Topaz Mountain, located at the southern end of the range). SM = Spor Mountain, mainly made up of westward-dipping Devonian dolomites and older rocks, faulted upward at the eastern side of the mountain. Be = area of most of the beryllium workings, located in the beryllium tuff at the base of the 21 Ma Spor Mountain Formation (black on the map). D = The Dell, the valley between Spor Mountain and the Thomas Range; it contains additional beryllium claims and the Yellow Chief uranium mine. AR = Antelope Ridge, a subsidiary flow complex of Topaz Mountain rhyolite with excellent exposures of surge deposits at its southwest border. Q = Quaternary alluvium.

the Topaz Mountain rhyolite, in marked contrast to that associated with the nearby Spor Mountain rhyolite.

The planar and cross-bedded structures in the pyroclastic deposits that host the Spor Mountain beryllium mineralization earlier led to their classification as "water-laid" tuff, an apellation which has been difficult to erase from the geologic literature on this area. Thick, massive beds of the Spor Mountain tuffs contain abundant carbonate clasts set in a fine-ash matrix, features which suggest to us emplacement as cold, wet surges and lahars (cf. Sheridan and Wohletz, 1983). The common occurrence of impact sags and other features of soft-sediment deformation imply near saturation of the deposit with water at the time of emplacement. There is no welding or other evidence of high-temperature emplacement; Plinian-fall deposits are absent. These features are compatible with hydromagmatic explosions triggered by the interaction of magma with water in shallow aquifers of the carbonate rocks.

The unmineralized tuffs that underlie the Topaz Mountain rhyolite, in contrast, mainly consist of Plinian-fall deposits, ash-flow beds, and dry-surge sequences (well exposed on the southwestern side of Antelope Ridge; AR on Fig. 1). Pumice is more abundant than lithic clasts, locally the tuffs are unaltered, and ash-flow beds near the top of the sequence are locally partly welded. Apparently, these tuffs were emplaced at relatively high temperatures, and hydromagmatism played little or no role in the eruption of this sequence.

At Spor Mountain, bertrandite, fluorite, amorphous silica, and Mn-Fe oxides replace dolomite fragments in tuffaceous surge deposits just beneath the topaz-bearing rhyolitic lava flow. The Be-mineralized zone is also highly enriched in F, Li, Mn, U, Sn, W, Nb (and presumably Ta, though we have no analyses), Zn, Pb, and several other metals (but not Mo, possibly due to a lack of sulfur to form molybdenite). It is not known what phases contain the metals other than beryllium (in bertrandite), lithium (in montmorillonite), manganese (in black oxides), and uranium (in oxides and uraniferous opal). The uniform lateral character of the mineralization, the restriction of beryllium mineralization to the uppermost few meters of tuff, and its absence in vertical fractures or in fluorite-bearing breccia pipes (tuffaceous vent breccias, in some cases) in underlying dolomite, all lead to the conclusion (Bikun, 1980), that mineralization may have resulted from the devitrification of the cooling, overlying lavas (a "steam iron" model). Hot gases released during devitrification penetrated the cooler, wet, calcareous (containing abundant dolomite fragments) pyroclastic beds and generated a mineralization front just beneath the lava/tuff contact. Bertrandite coprecipitated with fluorite and replaced the carbonate clasts in the tuff. Mass balance calculations based on comparisons of chemical compositions of glassy and devitrified rhyolite are consistent with this interpretation.

An alternate explanation for the mineralization (Staatz and Carr, 1964, and numerous later workers) invokes meteoric or magmatic fluids rising from the presumed vent areas and spreading outward in tuff beneath the capping rhyolite flow. These fluids, if meteoric, could be of local derivation from an aquifer, convecting in response to the heat of a shallow pluton, and the beryllium could be leached from this pluton or more locally from the tuffs. Hot-spring activity in the Spor Mountain area continued for many millions of years following the initial eruption, as shown by age dating of uraniferous opal (Ludwig and others, 1980). Nevertheless, using this near-surface hydrothermal-convection model, it is difficult to explain the lack of beryllium in the fluorite breccia pipes cutting dolomite and in intermediate and lower levels of the tuffs.

An interesting feature of the crystal-rich topaz rhyolite lava southwest of Spor Mountain is the common presence of quenched fragments of more mafic lava, each surrounded by a vapor-phase altered halo (Christiansen and others, 1981). Amoeboid pillows of dark material are locally flattened into layers that are complexly folded, and perhaps indicate a plastic condition during their emplacement. The implied mechanical mixing of the two magma types must have occurred very shortly before or during the eruption, rather than earlier, because the more mafic lava remained hot long enough to promote local vapor phase evolution in its rhyolite host as the two lava types reached thermal equilibrium *after* emplacement. (Some chemical mixing could have occurred earlier, in a compositionally stratified magma chamber.) This feature has not been reported from other topaz rhyolite lavas, and it is unknown if injection of the more mafic lava caused eruption of the highly evolved Spor Mountain rhyolite or if the mechanical mixing of the two magma types bears any relation to the Be mineralization that is unique to Spor Mountain.

The Yellow Chief uranium mine is located between Spor Mountain and Topaz Mountain, in a down-faulted area called the Dell (Fig. 1). The mineralization consists of yellow oxidized uranium minerals in clastic rocks beneath the Spor Mountain tuff; the uranium was presumably leached from the tuff by descending ground water (Lindsey, 1982).

Most topaz rhyolites of the western United States are not economically mineralized. This lack of economic mineralization could be due to various factors (Burt and others, 1982), many of which remain poorly understood. Possible factors include the degree of fractionation of the magma, the eruption and cooling history of the lavas and associated tuffs, the presence or absence of reactive carbonate rocks in the vent areas, and the presence or absence of suitable fluids to cause mineralization. At Spor Mountain, mineralization presumably resulted from the juxtaposition of (1) an extremely differentiated and fluorine-enriched, cooling rhyolite lava, (2) reactive carbonate rock in the vent areas, resulting in its presence in the tuff underlying the lava, and (3) fluids, whether magmatic or meteoric in origin, that caused alteration and mineralization at the top of the tuff layer.

MEXICAN-TYPE TIN DEPOSITS

Mexican-type fumarolic tin deposits occur along fractures toward the tops of rhyolitic domes. Lower temperature dissolution and reworking of earlier fumarolic deposits may actually produce the colloform "wood tin" typical of this deposit type (Correa, 1981). Red, higher temperature crystalline cassiterite, such as that described by Lufkin (1976), occurs more rarely. Deposits of this type occur in Nevada and New Mexico, as well as in many areas of central and northern Mexico (especially the states of Guanajuato, San Luis Potosi, Zacatecas, and Durango; Foshag and Fries, 1942). The domes in all places appear to be more than 20 m.y. old, and the ones that have been dated in Mexico cluster close to 30 Ma (Huspeni and others, 1984). This restricted age range, the abundance of tin occurrences, and the occurrences on a high plateau on the eastern side of the Cordillera (rather than a "torn-apart" province such as the Basin and Range or Rio Grande rift) appear to set the Mexican topaz rhyolites apart from those in the United States. Nevertheless, major rifting did occur in this region of Mexico, as revealed by the presence of numerous large graben structures, but apparently not over an extended time period.

While most, but not all (cf. Rye and others, 1984) researchers agree that the tin is of local derivation, released by the devitrification of the rhyolitic dome itself (cf. Duffield and others, 1984; Eggleston and Norman, 1984), there seems to be little

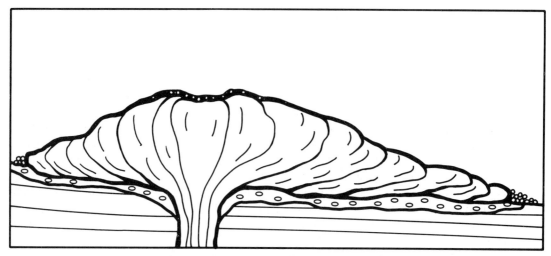

Figure 2. Hypothetical cross section through a rhyolite dome showing the steepening and overturning of flow foliations as the upper contact or carapace is approached.

agreement about the volcanologic controls on the mineralization, particularly in Mexico. Recent literature (Pan, 1974; Lufkin, 1977; Huspeni and others, 1984) suggests that overlying ignimbrite sheets (cap ignimbrite) may have played a role as an impermeable cap above a mineralizing system, and that the mineralization only occurs in breccias and vertically flow-banded rhyolite in vent areas; this presumed relation to vents has implied to some researchers, such as Rye and others (1984), a deeper origin for the tin.

Our field research in Mexico during 1984 has led us to an alternate explanation of the role of the capping ignimbrite and of domal structures in localizing tin mineralization, at least in numerous San Luis Potosi and Guanajuato occurrences. (Duffield and others, 1984, reached a similar conclusion for occurrences in Durango.) We conclude that the capping ignimbrite commonly found in the vicinity of tin occurrences merely plays a passive role in preserving these occurrences. (There is no capping ignimbrite in the Guanajuato tin rhyolite occurrences.) Where present, the ignimbrite is unmineralized, and served to protect the easily stripped, brecciated carapace of the domes from later erosion, until a time close to the present. Where the protection of the ignimbrite has only recently been removed, tin deposits remain in the carapace of the domes. Father from outcrops of the ignimbrite, the dome has typically been eroded more deeply, and tin, if it remains at all, is present only in placer deposits. The lack of such ignimbrites associated with most occurrences of topaz rhyolites in the United States might account, in part, for the lack of tin associated with most of them (it may have been eroded away). Economic factors, in terms of lack of incentive for prospecting, may also partly explain the paucity of identified U.S. occurrences.

What about the reported restriction (Pan, 1974) of tin deposits to vent breccias and related vertical flow banding? We conclude that this well-established "fact" is generally fiction.

Figure 2 shows a hypothetical cross section through a rhyolite dome that shows a possible pattern of flow banding. The important point is that flow banding commonly steepens or overturns as it nears the breccia carapace, even at extreme distances from the vent area (cf. similar observations on rhyolite by Christiansen and Lipman, 1966, and on pantellerite by Bryan, 1966, p. 467). The cause is presumably friction or drag on the cooler, more viscous carapace. We observed this pattern in numerous domes with related tin mineralization. We also saw the transition from near-vertical flow banding to carapace breccia to cap ignimbrite.

We observed that tin mineralization occurred in the following five environments (illustrated in Fig. 3): (1) in veins and fractures that cut across the flow banding, (2) in steeply flow-banded rhyolite, mineralization occurring along the flow banding, (3) commonly in vitrophyric carapace breccia, usually silicified and altered, (4) rarely in vitrophyric basal breccia, usually where the flow had bent up on meeting an obstacle to its progress, and (5) in placer deposits beyond the flow. Only the first mode of mineralization is exploited by underground mine workings; the other modes are mined from small surface pits and holes in which the tin can be washed from the soil. Carapace breccias (the third mode above) appear to be responsible for most production.

We hypothesize that topaz, when it occurs in these rhyolites, is formed in cavities at depths beneath those at which fumarolic tin is deposited (see Fig. 3). We found topaz typically in areas of nearly horizontal flow banding. This hypothesis could explain why topaz may or may not be reported from the tin-bearing rhyolites. In addition, many of the tin-bearing rhyolites probably have no topaz in vesicles at any depth. That is, there is not necessarily a direct connection between tin-bearing rhyolite domes and topaz-bearing rhyolite domes. The former are generally relatively large (several kiometres across); the latter can be exceedingly small (less than 0.5 km across).

Only in a very few places did we observe a tin working that

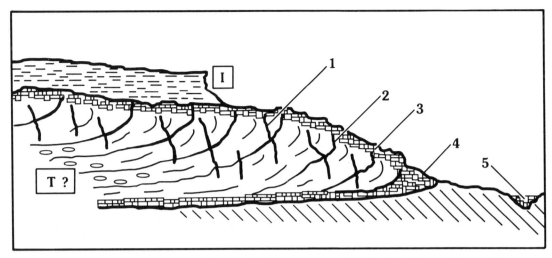

Figure 3. Schematic cross section through a Mexican-type topaz rhyolite flow showing five possible environments of tin mineralization and their erosional relation to a capping ignimbrite (if present): 1 = along internal fractures that cut flow banding; 2 = along flow-banding where it steepens toward the carapace; 3 = in the vitrophyric carapace breccia; 4 = in the breccia beneath and at the front of the flow; 5 = in placer deposits formed by erosion of the flow. I = the capping ignimbrite (commonly absent). T? = possible occurrence of topaz in miarolitic cavities within the flow, where flow banding is horizontal.

we interpreted to be in a vent area. Such areas (as inferred by dome morphology and concentric patterns of flow banding) are commonly covered by soil and vegetation and/or by ignimbrite. Tin workings typically were on moderate to steep slopes, characteristic of areas near the front of a rhyolite flow. Except in the Pinos area, Zacatecas, the base of the flows and any pyroclastic rocks underlying them are obscured by alluvium. Spor Mountain-type deposits, if present, could only be discovered by an expensive drilling program. A favorable prospecting criterion is the common presence of Mesozoic carbonate rocks in the preeruption basement.

The fact that tin workings were seen in a wide variety of rhyolite domes is strong evidence that the tin is of local fumarolic derivation, rather than from a hypothetical underlying pluton, as proposed by Rye and others (1984) for the Black Range, New Mexico. The latter interpretation would imply that each rhyolite dome is underlain by a pluton at just the right depth as well as by other structural conditions necessary for the deposition of tin in the overlying rhyolite dome.

INTRUSIVE EQUIVALENTS

Intrusive fluorine-rich rhyolitic laccoliths, plugs, or plutons may also host mineralization. Familiar examples include the well-known Climax-type or granite-type porphyry molybdenum deposits of Colorado, New Mexico, and Utah (cf. Mutschler and others, 1981; White and others, 1981; Westra and Keith, 1981), the prophyry tin deposits of Bolivia (Sillitoe and others, 1975; actually these appear to be enriched in boron rather than fluorine), and the porphyry tungsten deposit at Mount Pleasant, New Brunswick, Canada (Parrish and Tully, 1978). Of these, the Climax-type porphyry molybdenum deposits appear to be co-magmatic with western U.S. topaz rhyolites (Burt and others, 1982). Similarly, topaz and tourmaline-bearing granites that we have observed near Guadalcazar, San Luis Potosi, and at Peñon Blanco, Zacatecas, appear to be comagmatic with Mexican topaz rhyolites. The former area has been mined for tin and silver and the latter prospected for fluorite and uranium (information based on visit by Burt). Tourmaline, abundant in the two Mexican granites, is not a characteristic phase of United States porphyry molybdenum deposits or topaz-bearing granites.

Mineralization in these deposits typically is also adjacent to or within the former carapace or shell, and likewise appears to be magmatically derived from the vapor phase released during crystallization of the host. Topaz is a characteristic alteration mineral, except in the Bolivian tin deposits, which contain tourmaline. Although these deposits are subvolcanic or deeper, they have many features in common with the volcanic deposits discussed above, and might well be found beneath them (as apparently was the case at Climax, Colorado).

Base and precious metal deposits, rich in silver, may occur in veins peripheral to the plutons, and related epithermal gold or silver deposits might be found in the overlying volcanic rocks (including fluorine-rich rhyolite domes). The current consensus (cf. Buchanan, 1981; White, 1981) is that these deposits, commonly hosted by rhyolitic volcanic rocks, are formed by circulating meteoric fluids driven by deeper plutons.

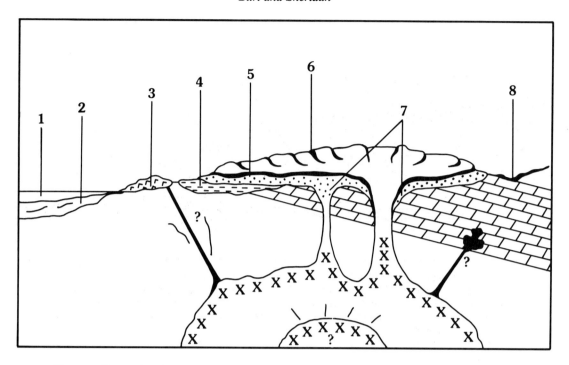

Figure 4. Types of surface and near-surface mineralization possibly associated with topaz rhyolite volcanism (modified from Burt and others, 1982, Fig. 4). (1–3 are speculative; 4–8 are observed.) 1 = brines (Li, W, B, etc.); 2 = tuffaceous lake sediments (U, Li, Mn, etc.); 3 = hot spring deposits (W, Mn, Ag, Au, etc.); 4 = clastic rocks beneath tuffs (U); 5 = mineralized pyroclastic deposits beneath flows (Be, F, Li, Cs, etc.); 6 = fractured and brecciated carapace of domes and flows (fumarolic Sn); 7 = vent and contact breccias (F, U, etc.); 8 = placer deposits (Sn, gemstones, etc.). Question marks indicate possible subvolcanic types of deposits, including veins (precious-metal type), replacements (skarn type), and disseminated bodies (porphyry type).

SUMMARY AND CONCLUSIONS

Many types of mineral deposits appear to be genetically related to fluorine-rich rhyolite volcanism. These are summarized in Figure 4 (cf. Burt and others, 1982, for a more complete discussion with examples). These include (1) Li and possibly W and B in playa-lake brines, (2) U, Li, and Mn in tuffaceous lake sediments, (3) U, Mn, W, Hg, Ag, and Au in hot springs deposits, (4) U in clastic rocks underlying tuffs (Yellow Chief Mine type), (5) Be, Sn, W, Li, Cs, and U in tuffs underlying topaz rhyolite flows (Spor Mountain type), (6) Sn in rhyolite domes and flows (Mexican type), (7) fluorite, uranium, and possibly precious metals in vent and hydrothermal breccias, and (8) placer deposits of tin and gemstones. Mineralization types 3, 4, 5, and 7 have been recognized to date in the Spor Mountain area; types 6 and 8 are found in association with Mexican tin rhyolites. Question marks in Figure 4 indicate possible subvolcanic vein, replacement, and disseminated (porphyry-style) deposits.

This review has concentrated on only three of the above types of deposits associated with volcanic-plutonic processes: (1) Spor Mountain-type deposits hosted by pyroclastic rocks just beneath a topaz rhyolite lava flow, (2) Mexican-type fumarolic tin deposits found in and just beneath the brecciated carapace of rhyolite domes and flows, and (3) Climax-type porphyry deposits found near the carapace of subvolcanic rhyolitic plutons. Although one could argue for a deeper, more distant source for all three types of mineralization, the evidence presented here suggests that all three are locally derived by devolatilization during crystallization of the immediately adjacent melt.

ACKNOWLEDGMENTS

Our initial research on the topaz rhyolites of the western United States was supported by U.S. Department of Energy Subcontract 79-270-E, and our later work in Mexico was supported by National Science Foundation Grant INT-8213108 and by a research grant from the Arizona State University College of Liberal Arts. We are grateful to the numerous personnel and institutions in Mexico that have cooperated with our research, especially Guillermo Labarthe and Alfredo Aguillon of the University of San Luis Potosi, Juan Randall of the University of Guanajuato, and Jorje Arranda of the Guanajuato branch of the Autonomous University of Mexico (UNAM). We also thank Eric Christiansen, Joaquin Ruiz, and Wendell Duffield for many fruitful discussions and for their careful reviews of an earlier version of this paper.

REFERENCES CITED

Bikun, J. V., 1980, Fluorine and lithophile element mineralization at Spor Mountain, Utah: U.S. Department of Energy Open-File Report, GJBX-225(80), p. 167–377.

Bryan, W. B., 1966, History and mechanism of eruption of soda-rhyolite and alkali basalt, Socorro Island, Mexico: Bulletin Volcanologique, v. 29, p. 453–479.

Buchanan, L. J., 1981, Precious metal deposits associated with volcanic environments in the southwest: Arizona Geological Society Digest, v. 14, p. 237–262.

Burt, D. M., and Sheridan, M. F., 1981, A model for the formation of uranium/lithophile element deposits in fluorine-rich volcanic rocks: American Association of Petroleum Geologists Studies in Geology, v. 13, p. 99–109.

Burt, D. M., Sheridan, M. F., Bikun, J. V., and Christiansen, E. H., 1982, Topaz rhyolites—Distribution, origin, and significance for exploration: Economic Geology, v. 77, p. 1818–1836.

Christiansen, E. H., Burt, D. M., and Sheridan, M. F., 1981, Evidence for magma mixing in rhyolites from Spor Mountain, Utah: Geological Society of America Abstracts with Programs, v. 13, p. 426.

Christiansen, E. H., Burt, D. M., Sheridan, M. F., and Wilson, R. T., 1983, Petrogenesis of topaz rhyolites from the western United States: Contributions to Mineralogy and Petrology, v. 83, p. 16–30.

Christiansen, E. H., Bikun, J. V., Sheridan, M. F., and Burt, D. M., 1984, Geochemical evolution of topaz rhyolites from the Thomas Range and Spor Mountain, Utah: American Mineralogist, v. 69, p. 223–236.

Christiansen, R. L., and Lipman, P. W., 1966, Emplacement and thermal history of a rhyolite lava flow near Fortymile Canyon, southern Nevada: Geological Society of America Bulletin, v. 77, p. 671–684.

Correa, B. P., 1981, The Taylor Creek rhyolite and associated tin deposits, southwestern New Mexico [M.S. thesis]: Tempe, Arizona State Univresity, 104 p.

Dingwell, D. B., 1985, Viscosities in the system Albite-H_2O-F_2O_{-1} [abs.]: EOS (American Geophysical Union Transactions), v. 66, p. 392.

Dingwell, D. B., Scarfe, C. M., and Cronin, D. J., 1985, The effect of fluorine on viscosities in the system Na_2O-Al_2O_3-SiO_2: Implications for phonolites, trachytes, and rhyolites: American Mineralogist, v. 70, p. 80–87.

Duffield, W. A., Ludington, S. D., Maxwell, C. H., Reed, B. L., and Richter, D. H., 1984, Tin mineralization in rhyolite lavas, Sierra Madre Occidental, Mexico: Geological Society of America Abstracts with Programs, v. 16, p. 495.

Eggleston, T. L., and Norman, D. I., 1984, Geochemistry and origin of rhyolite-hosted tin deposits, southwestern New Mexico: Geological Society of America Abstracts with Programs, v. 16, p. 499.

Foshag, W. F., and Fries, C., Jr., 1942, Tin deposits of the Republic of Mexico: U.S. Geological Survey Bulletin 935-C, p. 99–176.

Huspeni, J. R., Kesler, S. E., Ruiz, J., Tuta, Z., Sutter, J. F., and Jones, L. M., 1984, Petrology and geochemistry of rhyolites associated with tin mineralization in northern Mexico: Economic Geology, v. 79, p. 87–105.

Lindsey, D. A., 1977, Epithermal beryllium deposits in water-laid tuff, western Utah: Economic Geology, v. 72, p. 219–232.

—— , 1979, Geologic map and cross sections of Tertiary rocks in the Thomas Range and northern Drum Mountains, Juab County, Utah: U.S. Geological Survey Miscellaneous Geological Investigations Map I-1176, scale 1:62,500.

—— , 1982, Tertiary volcanic rocks and uranium in the Thomas Range and northern Drum Mountains, Juab County, Utah: U.S. Geological Survey Professional Paper 1221, 71 p.

Ludwig, K. R., Lindsey, D. A., Zielinski, R. A., and Simmons, K. R., 1980, U-Pb ages of uraniferous opals and implications for the history of beryllium, fluorine, and uranium mineralization at Spor Mountain, Utah: Earth and Planetary Science Letters, v. 46, p. 221–232.

Lufkin, J. L., 1976, Oxide minerals in miarolitic rhyolite, Black Range, New Mexico: American Mineralogist, v. 61, p. 425–430.

—— , 1977, Chemistry and mineralogy of wood tin, Black Range, New Mexico: American Mineralogist, v. 62, p. 100–106.

Manning, D.A.C., 1981, The effect of fluorine on liquidus phase relationships in the system Qz-Ab-Or with excess water at 1 kb: Contributions to Mineralogy and Petrology, v. 76, p. 206–215.

Mutschler, F. E., Wright, E. G., Ludington, S., and Abbott, J. E., 1981, Granite molybdenum systems: Economic Geology, v. 76, p. 874–897.

Pan, Y.-S., 1974, The genesis of the Mexican type tin deposits in acidic volcanics [Ph.D. thesis]: New York, Columbia University, 286 p.

Parrish, I. S., and Tully, J. V., 1978, Porphyry tungsten zones at Mt. Pleasant, N.B.: Canadian Institute of Mining and Metallurgy Bulletin, v. 71 (794), p. 93–100.

Rye, R. O., Lufkin, J. L., and Wasserman, M. D., 1984, Genesis of tin occurrences in the Black Range, New Mexico, as indicated by oxygen isotope studies: Geological Society of America Abstracts with Programs, v. 16, p. 642.

Sheridan, M. F., and Wohletz, K. H., 1983, Hydrovolcanism: Basic considerations and review: Journal of Volcanology and Geothermal Research, v. 17, p. 1–29.

Sillitoe, R. H., Halls, C., and Grant, J. N., 1975, Porphyry tin deposits in Bolivia: Economic Geology, v. 70, p. 913–927.

Staatz, M. H., and Carr, W. J., 1964, Geology and mineral deposits of the Thomas and Dugway ranges, Juab and Tooele counties, Utah: U.S. Geological Survey Professional Paper 415, 188 p.

Westra, G., and Keith, S. B., 1981, Classification and genesis of stockwork molybdenum deposits: Economic Geology, v. 76, p. 844–873.

White, D. E., 1981, Active geothermal systems and hydrothermal ore deposits: Economic Geology, 75th Anniversary Volume, p. 392–423.

White, W. H., Bookstrom, A. A., Kamilli, R. J., Ganster, M. W., Smith, R. P., Ranta, D. E., and Steininger, R. C., 1981, Character and genesis of Climax-type molybdenum deposits: Economic Geology, 75th Anniversary Volume, p. 270–316.

Wyllie, P. J., 1979, Magmas and volatile components: American Mineralogist, v. 64, p. 469–500.

Manuscript Accepted by the Society May 5, 1986

Printed in U.S.A.

An extensive, hot, vapor-charged rhyodacite flow, Baja California, Mexico

Brian P. Hausback
Geology Department
California State University
Sacramento, California 95819

ABSTRACT

The Providencia rhyodacite lava flow of southern Baja California is an unusually extensive salic extrusion. Remnants of the flow overlie lower to middle Miocene volcanic rocks and occur in a 27-km-long belt near the city of La Paz. Isopachs of the flow show a maximum thickness of 120 m and indicate a minimum volume of 8.6 km^3. Persistent flow bands are closely spaced and parallel the base of the flow. These flow bands are thin, planar lithophysal cavities that give the rock a distinct parting. In the upper part of the flow the banding is strongly deformed into isoclinal to open folds. Flow directions, developed from fold axial information, together with the isopach data, suggest that the rhyodacite flowed at least 23 km north-northwest from its source south of La Paz.

The Providencia rhyodacite (68–72.5% SiO_2, 3.8% Na_2O, and 4.5% K_2O) contains about 5% phenocrysts (plag > opx > Fe-Ti oxides) set in a devitrified groundmass of fine-grained alkali feldspar and tridymite ?). Lithophysal planar cavities are lined with large (as long as 3 mm) vapor-phase crystals whose paragenetic relationships define a crystallization order from oldest to youngest: (1) fayalite + thick laths of brown hornblende, (2) α-quartz, (3) hematite, and (4) tridymite + apatite + rare biotite + rare fibrous green hornblende.

Field evidence suggests, but does not prove, that the Providencia rhyodacite is a primary lava flow rather than a remobilized pyroclastic flow. A high volatile content together with a high eruption temperature acted in concert to maintain a low viscosity, a fact that probably facilitated flow of the lava to great distances.

INTRODUCTION

The Providencia rhyodacite flow of southern Baja California, Mexico (Fig. 1) is an important and possibly unique example of what appears to be a siliceous lava that flowed an unusually long distance from its source. Extrusions of rhyolite and rhyodacite typically form steep-sided, short, blocky flows—called coulees—such as the well-preserved Holocene glass flows found in the Medicine Lake Highland and at the Mono and Inyo craters, California. Lavas of these compositions rarely, if ever, flow more than 5 km from their eruptive vents; their morphology and size have been summarized by Walker (1973) and Clough and others (1982). Low fluidity and rapid vesiculation of dissolved gases from siliceous magmas commonly lead to explosive, large-volume eruptions and the resultant generation of ash-flow tuffs. Many voluminous, siliceous, lava-like deposits have been recognized as densely welded ash flows in which secondary mass flowage has occurred. The purpose of this paper is to document the occurrence, flow characteristics, and distinctive vapor-phase mineralogy of this unit and to investigate the possibility of its origin as a volatile-rich ignimbrite.

STRATIGRAPHY

In the La Paz area, the Providencia rhyodacite is the uppermost member of the early to middle Miocene Comondú Formation. This subaerially deposited calc-alkaline volcanic sequence predates the late Miocene to Recent opening of the Gulf of California and consists of silicic ignimbrites, andesitic lahars, and related volcaniclastic sandstones. The source of most of these volcanic units was a chain of volcanoes along the eastern margin of what is now the southern Baja California peninsula. The post-Comondú Late Tertiary rifting of the southern Gulf of California took place along the axis of the previously active Comondú continental volcanic arc (Hausback, 1984a). K-Ar radiometric age determinations on the Comondú Formation in Baja California Sur have yielded an age range of 25–12.5 Ma. The flow is newly named for exposures near Rancho la Divina Providencia near the city of La Paz and is quarried for a well-known resistant building stone. Plagioclase and glass separates from the Providencia lava from 3 widely separated exposures yield K-Ar ages of 19.1 ±1.2 Ma, 19.2 ±0.5 Ma, and 19.7 ±0.2 Ma. (Hausback, 1984a).

FIELD CHARACTERISTICS

The Providencia rhyodacite flow is a prominent mesa-capping unit exposed in a series of hills extending 27 km to the north and south of La Paz (Hausback, 1983). It is a medium-gray, resistant, dense, strongly flow-banded rock. Lack of internal contacts in all exposed sections suggests that the lava is a single flow unit. Identical field characteristics and stratigraphic position, as well as similar compositions (Table 1) and radiometric ages between the various isolated outcrops verify that they constitute a single flow.

The Providencia-flow-capped hills are generally flat-topped, but no original lava surface features such as levees or ogives were recognized; lack of these features suggests that the upper portion of the flow has been eroded. Measured stratigraphic sections were used to construct an isopach map (Fig. 2) from which a minimal preerosional volume of 8.6 km^3 was calculated. The unit is extremely resistant and overlies a 110-m-thick section of friable volcaniclastic sandstone and tuff (Fig. 3). Erosion of these less-competent units necessarily undermines the Providencia layer and leaves it as isolated cap rocks. The regular thickness decrease to the north and the flat-topped morphology of the outcrops also suggests that the flow remnants are eroded down to a natural stratigraphic zonal break. Of the original areal coverage, estimated at about 200 km^2, only 6% remains today.

The base of the flow shows only minor local relief, so that the unit was probably emplaced on a topographically smooth substrate. The isopachs indicate a maximum thickness of about 120 m toward the southern end of the flow exposures; this suggests that this area, known as Cerro San Ramon, is the site of the main eruptive vent for the flow. Therefore, the flow moved at least 23 km (of its 27 km extent) to the north from its source. The stratigraphy has been faulted and warped since deposition, so that the northernmost, distal exposure of the Prodivencia is now 250 m in elevation above its source area, 23 km to the south.

The base of the flow is rarely exposed due to large talus accumulations on the flanks of the mesas. However, on the east side of Cerro Calavera the flow is in sharp contact with an underlying, poorly exposed, thin, red-brown (baked?) ash-flow

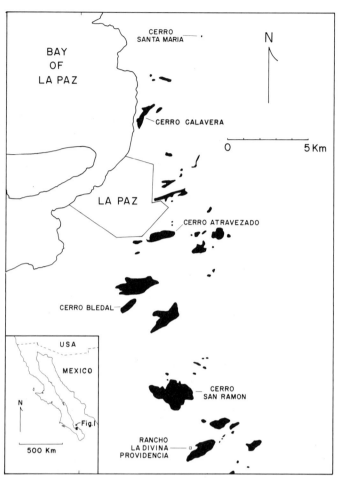

Figure 1. Index map of the La Paz area, southern Baja California, Mexico. Shaded areas indicate the distribution of the Providencia rhyodacite.

tuff. No tephra associated with the Providencia flow have been recognized along its base. The lower 5 m of the Providencia flow is a poorly sorted autobreccia that consists of angular boulders of black obsidian. The obsidian is generally massive and locally spherulitically devitrified. It is locally interlaminated with septa of pervasively devitrified laminated rock similar to the bulk flow above. Some obsidian blocks are highly vesicular. This obsidian zone almost surely represents the accumulation of blocks that fell from the advancing flow front, and the zone was subsequently covered by the flow itself.

The most striking feature of the Providencia rhyodacite is a uniquitous banding defined by thin (as much as 3 mm) white to light-gray laminations set in a medium-gray to brown groundmass (Fig. 4). Vapor-phase minerals wholly or partially encrust these planar laminations. The bands are laterally continuous (over several meters), planar lithophysal laminations that roughly parallel the base of the flow and are irregularly spaced from 2 to 10 mm apart. They are strongly deformed into isoclinal folds at the base of the lava and have horizontal axial planes. Toward the

TABLE 1. PROVIDENCIA RHYODACITE CHEMICAL ANALYSES

Specimen Wt %	433 (1)	432 dark (2)	432 light (2)	113M (2)	3 (2)	44 (2)	239 (2)	248 (2)	383 (3)	249 (3)	240 (3)	XRF ave. σ	383 (4)	383 (5)
SiO2	62.11	61.07	68.83	70.96	72.57	73.48	72.58	72.47	67.93	69.07	69.89	1.00	0.06	0
TiO2	1.19	1.21	0.58	0.48	0.47	0.42	0.48	0.47	0.46	0.47	0.48	0.05	18.64	49.23
Al2O3	16.28	15.95	13.65	13.48	13.70	13.73	13.62	13.88	12.79	13.10	13.28	0.90	1.42	0.03
FeO*	5.94	6.12	3.09	2.59	2.61	2.31	2.55	2.63	2.56	2.50	2.66	0.20	75.46	47.92
MnO	0.13	0.08	0.03	0.05	0.06	0.06	0.05	0.05	0.07	0.07	0.07	0.02	0.65	0.95
MgO	0.73	0.70	0.26	0.08	0.08	0.08	0.11	0.12	0.08	0.10	0.17	0.20	1.26	1.88
CaO	4.65	4.65	1.46	1.10	1.01	0.96	1.12	1.21	1.32	1.26	1.37	0.15	0	0.01
Na2O	3.95	4.08	3.83	3.77	3.87	3.96	3.82	4.03	3.49	3.92	4.26	0.30	----	----
K2O	3.12	2.83	4.39	4.58	4.58	4.92	4.56	4.74	4.61	4.59	4.10	0.50	----	----
V2O5	----	----	----	----	----	----	----	----	----	----	----	----	0.17	0
Cl (ppm)	38	49	39	38	96	72	88	148	717	560	693		----	----
Ba (ppm)	1061	1160	1651	1653	1502	1453	1558	1364	1361	1427	1537		----	----
L.O.I.	----	----	----	----	----	----	----	----	3.86	----	----		----	----
Total	98.21	96.81	96.30	97.24	99.12	100.07	99.06	99.75	97.38	95.27	96.50		97.66	100.02

(1) Intermingled andesite
(2) bulk rhyodacite
(3) basal obsidian
(4) magnetite-ulvospinel
(5) ilmenite-hematite

Analyses 1, 2, and 3 from XRF.
Analyses 4 and 5 from electron microscope.

*Total iron reported as FeO.
L.O.I. = Loss on ignition.

top of the exposures, these folds become progressively more open, less inclined, and attain amplitudes of up to 5 m.

Closely spaced minute parasitic folds with amplitudes up to 3 mm are superimposed on the larger folds throughout the upper parts of the flow. The axes of the parasitic folds are nearly always parallel to the axes of the larger folds. Locally, however, these subordinate structures are refolded by their larger hosts. Figure 2 shows stereographic plots of the average orientations of fold axes from several outcrops throughout the La Paz area. The fold axes in any one outcrop area show a wide scatter in orientation (Fig. 5), as would be expected for deformational features of this type formed by internal body forces within a viscous flowing medium.

After each fold was formed, it was subjected to rotation in the direction of continued shear, as are all linear structural elements in a flow. Nonetheless, the fold axes are not randomly oriented, and the mean orientation of the fold axes is the best available estimate of the original attitude of the folds. The average fold axis in any one area is subhorizontal with a trend consistently perpendicular to the northerly flow direction, as shown in Figure 2. This results from a process which is analogous to a carpet sliding down a ramp, folding and rolling over itself under its own weight, the fold axes developing perpendicular to the direction of movement. Fold vergence and rolling directions of andesitic magma blebs within the flow (Fig. 6) give the sense of movement to otherwise dipolar flow directions derived from the mean fold axes. The average orientations of the flow folds, together with the isopach data, indicate that the flow moved almost due north from its source at Cerro San Ramon.

PETROLOGY

The Providencia flow is a medium-gray, phenocryst-poor rhyodacite. The groundmass is generally nonvesicular and devitrified to alkali feldspars, mainly sanidine, and some form of very fine-grained silica, probably tridymite, based on X-ray powder diffraction and electron microprobe analyses. An exception to this is the black, vesicular, microlitic obsidian found only in a boulder breccia along the base of the flow in all known exposures.

The flow contains about 5% phenocrysts: plagioclase (An = 30–42% with oscillatory to slightly reversed zoning) > hypersthene > Fe-Ti oxides. The assemblage is relatively homogeneous throughout the flow. At the suspected source area, however, the flow is composed of a mixed lithology; the typical gray phenocryst-poor rhyodacite is intermingled with formless blebs and elongate swirled septa of dark-brown andesite (Fig. 7) that contains 12% phenocrysts of a similar assemblage, but with augite instead of hypersthene. Contacts between the two lithologies are marked by lithophysal laminations. Neither lithology appears quenched or in any way affected along the contacts. The andesite was probably liquid when it was incorporated into the rhyodacite, unlike other scarce, accidental, angular clasts of andesite in the flow.

XRF whole-rock analyses (Table 1) show that the lava flow is a metaluminous, alkaline rhyodacite to rhyolite. The compositions of the bulk flow and the basal obsidian determined from numerous samples along the length of the flow show almost no significant variation, with the exception of SiO_2. SiO_2 ranges

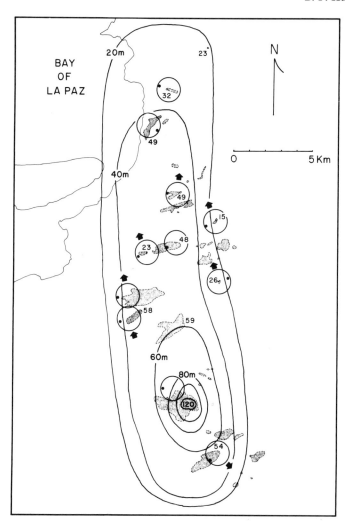

Figure 2. Isopach reconstruction map and flow-fold information. A flow volume of 8.6 km³ is calculated within the 20-m isopach. Small numbers are the flow thickness in meters. Circles are lower hemisphere, equal-area, oriented stereograms; dots within circles indicate the average fold-axis orientation; they yield a nonunique flow direction perpendicular to the fold axis. The arrows show the inferred sense of flow (determined from fold vergence and rolling of andesitic magma blebs).

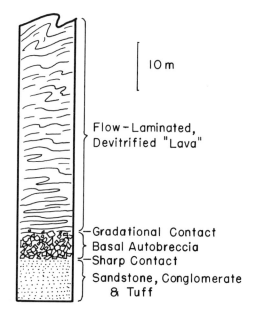

Figure 3. Composite stratigraphic section of the Providencia rhyodacite.

from 68–72.5%, a variation that may result from magmatic differentiation prior to eruption or from the remobilization and redistribution of SiO_2 during posteruptive vapor escape. The andesite lithology at the source area represents a mafic liquid that either intruded into the base of the magma chamber, an event which may have caused the eruption of the Providencia lava, or was a coexisting phase of magma that intermingled and erupted with the last-erupted rhyodacite.

Fe-Ti oxide geothermometry (Buddington and Lindsley, 1964; Carmichael, 1967; Ghiorso and Carmichael, 1981) indicates an eruptive temperature of 912 °C and an oxygen fugacity of $10^{-12.2}$, on the basis of microprobe analyses of fresh, unexsolved Fe-Ti oxide crystals from the quenched basal obsidian (Table 1). Eruption temperatures for continental rhyolites and rhyodacites have an approximate range of 790–925 °C (Carmichael and others, 1974), implying that the Providencia rhyodacite was relatively hot when erupted.

VAPOR-PHASE CRYSTALLIZATION

The planar lithophysal partings, so characteristic of the flow, are incompletely filled by a remarkable and diverse suite of vapor-precipitated euhedral crystals (up to 3 mm in size). This mineralization, although commonly overlooked in volcanic rocks, is of great importance in understanding the large volcanic vapor component and its effect on flow rheology. These vapors, except where incorporated into these minerals, are forever lost. The minerals are found throughout the flow, both laterally and vertically through most sections. The vapor phase minerals comprise 1–5% of the total volume of the Providencia flow. These euhedral crystals have precipitated as radiating clusters on the walls of the planar cavities and normally fill the openings.

Eight different mineral phases have been identified in the lithophysae by optical, X-ray diffraction, and electron-microprobe analyses. Paragenetic relationships define a crystallization sequence from oldest to youngest: (1) fayalite + thick laths of brown hornblende (with 3.2% F, 0.25% Cl, and probably containing H_2O); (2) α-quartz; (3) acicular hematite; and (4) tridymite + apatite (with 5% F and probably containing CO_2) + rare biotite + rare fibrous green hornblende (Hausback, 1984b). This unusual suite of vapor-phase minerals is the subject of a separate topical study (Hausback, in prep.).

Figure 4. Typical outcrop of the Providencia rhyodacite displays dense, resistant lava with closely spaced subhorizontal lithophysal laminations. Inclusions near the hammer head are intermingled andesite. Hammer head is 20 cm long.

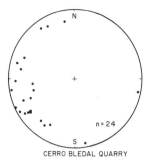
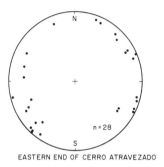

Figure 5. Fold axis orientations from two localities of the Providencia rhyodacite, plotted on lower hemisphere equal-area stereonets.

The planar lithophysal cavities in which this entire crystal suite is developed are laterally continuous over several meters and wrap around magmatic phenocrysts without breaking them. This fact suggests that movement along these bands occurred while the flow was still molten or in a plastic state. As such, the bands were probably shear planes active in the last stages of laminar flow in the cooling mass. Within the laminations, the vapor phase crystals also show no breakage, strongly indicating that they developed *after* all movement along the flow shears ceased. The ubiquitous postemplacement vapor-phase mineralization found in the Providencia flow strongly suggests that an enormous amount of vapor was dissolved in the magma and was, at least partially, retained in solution until the flow settled. The compositions of the vapor-phase crystals show that the lost volatiles of the voluminous volcanic vapor included H_2O, CO_2, F_2, and Cl_2.

Water and CO_2 are the most abundant volatile components commonly found in magmatic vapors (Burnham, 1979), and there is no reason to believe that the vapors associated with the Providencia rhyodacite flow were dissimilar. Primary hydrous phenocrysts such as biotite or hornblende would be expected in a volcanic rock of this composition, but there are none. This can be explained by the calculated high eruptive temperature, which precludes hydrous phenocryst stability, according to the experimental results of Maaloe and Wyllie (1975) and Naney (1983).

SUMMARY AND DISCUSSION

The Providencia rhyodacite is a remarkably long (at least 23 km) silicic volcanic flow that appears to be a lava. The exact original morphology of the flow cannot be accurately reconstructed due to extensive lateral and unconstrained vertical erosional removal since its emplacement in the early to middle Miocene. However, a reasonable reconstruction of this flow (Fig. 2) shows that it is far greater in length than other commonly known siliceous lavas.

The length of a lava flow is controlled by numerous factors, including lava viscosity, total erupted volume, rate of effusion, and morphology and slope of the underlying surface. Walker (1973) suggested that the length of a lava flow is most dependent on eruption rate and viscosity. Malin (1980), on the other hand, suggested that length depends most directly on the total volume of lava erupted. Unfortunately, these studies concentrated on mafic lavas. Very few silicic lavas have ever been observed in eruption, hence in situ viscosity and rates of eruption for salic lavas are essentially unknown.

The calculated Providencia flow volume of 8.6 km^3 or more is large in comparison to most salic lavas. Glass Mountain, probably the largest, best-known Quaternary rhyolite lava, is about 1 km^3 in volume (Eichelberger, 1981) and flowed 2.5 km down steep topography from its vent. However, much larger silicic lavas exist. The Chao dacite flow in northern Chile is 24 km^3 (Guest and Sánchez, 1969) and flowed 12 km from its vent. It is quite clear from their well-preserved surface morphology that young flows like these are primary lavas. However, old and poorly preserved flows such as the Providencia rhyodacite have less obvious interpretations; they possibly originated as either primary lavas or remobilized pyroclastic flows. Both of these eruptive styles commonly give rise to flow-banded rhyolitic extrusions.

Figure 6. Photographs showing (a) fold vergence (example from a loose block) and (b) rolling of an andesitic magma bleb. Both indicate flow to the left.

Field evidence argues against an ash-flow origin for the Providencia rhyodacite and suggests that it is actually a primary lava. There is no evidence for an explosive eruption that commonly gives rise to pyroclastic flows. No basal tephra unit has been found associated with the Providencia flow. In addition, the scarcity of accidental rock fragments in the flow, a typically abundant constituent of ash-flow tuffs, argues against an explosive pyroclastic origin (R. L. Smith, 1985, personal commun.). Furthermore, pyroclastic textures on the microscopic and megascopic scales are completely absent from the Providencia flow, whereas rheoignimbrites reported in the literature (Wolff and Wright, 1981; Chapin and Lowell, 1979; Deal, 1973; Parker, 1972; Lock, 1972; Villari, 1969; Noble, 1968; Walker and Swanson, 1968; Schmincke and Swanson, 1967) show, at least locally, remnant pyroclastic textures.

There is no doubt that the Providencia rhyodacite last flowed as a lava. Vesicular obsidian blocks in the basal zone show that it was a bubbling liquid mass. The foliation, folding, and vertical zonation of the lava are nearly indistinguishable from other well-documented salic lava flows, such as the flow described by Christiansen and Lipman (1966).

Viscosity indirectly controls the length of lava flows (Walker, 1973). The viscosity of the Providencia rhyodacite can be calculated by the method of Shaw (1972). By using this technique, the anhydrous viscosity of the Providencia lava is determined to be 10^8 Pascal-seconds (10^9 poise), assuming the derived eruptive temperature of 912°C. This estimate suggests that the Providencia flow, *if it was anhydrous,* was relatively fluid in comparison to less siliceous dacitic lavas that have measured viscosities in the range of 10^8–10^{10} Pa-s (10^9–10^{11} poise) (Walker, 1973).

The chemistry and amount of volatiles in solution during the eruption and flow of the Providencia rhyodacite is qualitatively suggested by the voluminous and ubiquitous postemplacement

Figure 7. Hand sample of intermingled rock types from the suspected vent-source locality. Swirl is dark andesite within lighter gray rhyodacite. Sample is 12 cm across in the horizontal dimension.

vapor-phase mineralization. The vapor component must have been substantial, and included at least H_2O, CO_2, F_2, and Cl_2.

The volatile component of lavas is known to significantly lower silicate fluid viscosity. Shaw (1963) showed that the addition of 7% H_2O to a rhyolitic glass can lower its viscosity by a factor of 10^7. The effects of F_2 are also important in volcanic rocks. F_2 and H_2O have similar effects in lowering silicate melt viscosity. In addition, F_2 has a high melt/fluid partitioning coefficient and will therefore be retained in a magma upon eruption, thereby reducing the explosivity of an eruption and generally maintaining the fluidizing effect of the F_2 in volcanic flows (Dingwell and others, 1985; Rabinovich, 1983).

The original volatile content of the Providencia lava is difficult to estimate. A glass separate from the obsidian basal breccia yielded a volatile content (retained in the glass at 110 °C) of 3.67 wt.%. If it is assumed that this 3.67% volatile content represents the original $H_2O + F_2$ content, the viscosity was again calculated by Shaw's (1972) method to be 10^4 Pa-s (10^5 poise), a value characteristic of much more fluid lavas such as andesites and basaltic andesites, which commonly *do* flow long distances (Macdonald, 1972; Walker, 1973). Therefore, it is reasonable that the relatively high temperature (912 °C) and a high volatile content combined to maintain a fluid Providencia lava, a quality that probably allowed it to flow such a great distance.

These arguments strongly suggest but do not unequivocally prove, that the Providencia flow was a primary lava. The preservation of a pyroclastic texture can prove a lava's remobilization from an ash flow, but lack of that texture cannot absolutely prove an origin as a primary lava. If a very hot, still-fluid ash flow is remobilized, it may strongly resemble a lava. If the pyroclastic texture becomes obscured through intense welding and laminar flowage, evidence of the primary eruptive character of the flow could be lost, but generally seems to be partially preserved. Indeed, there appears to be debate about the very nature of Rittman's (1958) type example of a rheoignimbrite (Wolff and Wright, 1981)—is it, in fact, a lava or a remobilized ash flow? Unusually extensive lavas such as the Providencia may always be suspected of having begun as ash-flow tuffs.

ACKNOWLEDGMENTS

Many people have given much toward the preparation of this manuscript. I am especially grateful to Adolf Pabst, who has been a constant source of inspiration, enthusiasm, and guidance throughout this study. In addition, Dr. Pabst has carried out many of the X-ray and optical experiments that have helped in the determination of the exact mineralogic character of several of the vapor-phase minerals.

Many others have been of great aid to me in this project. Joaquim Hampel and Howard Schorn helped solve numerous technical and experimental problems. M. Clark Blake originally suggested an in-depth look at this anomalous lava. I would also like to thank the many people who have given of their time to discuss and critique my written word and who have added much to the ideas presented here: James D. O'Brient, Jamie N. Gardner, Garniss Curtis, Clyde Wahrhaftig, Robert L. Smith, Bill Chávez, Mary Gilzean, Scott Linneman, Al Deino, Bill Murphy, and David Sussman.

I thank the Consejo de Recursos Minerales, especially Guillermo P. Salas, Salvador Casarrubias, Hugo Cortés, and Antonio Orozco; I also thank the Roca Fosfórica Mexicana, especially Francisco J. Escandón, Alfonso Martinez, and Carlos Perez Chacón for their generous financial and logistical support of this project.

REFERENCES CITED

Buddington, A. F., and Lindsley, D. H., 1964, Iron-titanium oxide minerals and synthetic equivalents: Journal of Petrology, v. 5, p. 310–357.

Burnham, C. W., 1979, The importance of volatile constituents, *in* Yoder, H. S., ed., The evolution of the igneous rocks: Princeton, New Jersey, Princeton University Press, p. 439–482.

Carmichael, I.S.E., 1967, The iron-titanium oxides of salic volcanic rocks and their associated ferromagnesian silicates: Contributions to Mineralogy and Petrology, v. 14, p. 36–64.

Carmichael, I.S.E., Turner, F. J., and Verhoogen, J., 1974, Igneous petrology: San Francisco, McGraw-Hill, Inc., 739 p.

Chapin, C. E., and Lowell, G. R., 1979, Primary and secondary flow structures in ash-flow tuffs of the Gribbles Run paleovalley, central Colorado, *in* C. E., and Elston, W. E., eds., Ash-flow tuffs: Geological Society of America Special Paper 180, p. 137–154.

Christiansen, R. L., and Lipman, P. W., 1966, Emplacement and thermal history of a rhyolite lava flow near Fortymile Canyon, southern Nevada: Geological Society of America Bulletin, v. 77, p. 671–684.

Clough, B. J., Wright, J. V., and Walker, G.P.L., 1982, Morphology and dimensions of the young comendite lavas of La Primavera volcano, Mexico: Geological Magazine, v. 119 (5), p. 477–485.

Deal, E. G., 1973, Primary laminar flow structures in the rhyolite ash-flow tuff of A. L. Peak, San Mateo Mountains, New Mexico: Geological Society of America Abstracts with Programs, v. 5, p. 475.

Dingwell, D. B., Christopher, M. S., and Cronin, D. J., 1985, The effect of fluorine on viscosities in the system $Na_2O-Al_2O_3-SiO_2$: Implications for phonolited, trachyted and rhyolites: American Mineralogist, v. 70, p. 80–87.

Eichelberger, J. C., 1981, Mechanism of magma mixing at Glass Mountain Medicine Lake Highland volcano, California, *in* Johnson, D. A., and Donnelly-Nolan, J., eds., Guides to some volcanic terranes in Washington, Idaho, Oregon, and northern California: U.S. Geological Survey Circular 838, p. 183–189.

Ghiorso, M. S., and Carmichael, I.S.E., 1981, A Fortran IV computer program for evaluating temperatures and oxygen fugacities from the compositions of coexisting iron-titanium oxides: Computers and Geosciences, v. 7, p. 123–129.

Guest, J. E., and Sánchez, R. J., 1969, A large dacitic lava flow in northern Chile: Bulletin Volcanologique, v. 33, p. 778–790.

Hausback, B. P., 1983, An extensive volatile-charged rhyodacite flow, Baja California, Mexico: Geological Society of America Abstracts with Programs, v. 15, p. 281.

——, 1984a, Cenozoic volcanic and tectonic evolution of Baja California Sur, Mexico, *in* Frizzell, V. A., Jr., ed., Geology of the Baja California peninsula: Society of Economic Paleontologists and Mineralogists, Pacific Section, v. 39, p. 219–236.

——, 1984b, Cenozoic volcanic and tectonic evolution of Baja California Sur, Mexico [Ph.D. thesis]: Berkeley, University of California, 162 p.

Lock, B. E., 1972, A lower Paleozoic rheo-ignimbrite from White Bay, Newfoundland: Canadian Journal of Earth Sciences, v. 9, p. 1495–1503.

Maaloe, S., and Wyllie, P. J., 1975, Water content of a granite magma deduced from the sequence of crystallization determined experimentally with water-undersaturated conditions: Contributions to Mineralogy and Petrology, v. 52, p. 175–191.

Macdonald, G. A., 1972, Volcanoes: Englewood Cliffs, New Jersey, Prentice-Hall, Inc., 510 p.

Malin, M. C., 1980, Length of Hawaiian lava flows: Geology, v. 8, p. 306–308.

Naney, M. T., 1983, Phase equilibria of rock-forming ferromagnesian silicates in granitic systems: American Journal of Science, v. 283, p. 993–1033.

Noble, D. C., 1968, Laminar viscous flowage structures in ash-flow tuffs from Gran Canaria, Canary Islands: A discussion: Journal of Geology, v. 76, p. 721–723.

Parker, D. F., 1972, Laminar flow in a peralkaline ash-flow sheet, Davis Mountains, Texas: Geological Society of America Abstracts with Programs, v. 4, p. 288.

Rabinovich, E. M., 1983, On the structural role of fluorine in silicate glasses: Physics and Chemistry of Glasses, v. 24 (2), p. 54–56.

Rittman, A., 1958, Cenni sulle colate di ignimbriti: Catania Accademia Gioenia di Scienze Naturali Bolletino ser. 4, p. 524–533.

Schmincke, H., and Swanson, D. A., 1967, Laminar siscous flowage structures in ash-flow tuffs from Gran Canaria, Canary Islands: Journal of Geology, v. 75, p. 641–664.

Shaw, H. R., 1963, Obsidian-H_2O viscosities at 1000 and 2000 bars in temperature range 700° to 900°C: Journal of Geophysical Research, v. 68, p. 6337–6343.

—— , 1972, Viscosities of magmatic silicate liquids: An empiracal method of prediction: Americal Journal of Science, v. 272, p. 870–893.

Villari, L., 1969, On particular ignimbrites of the island of Pantelleria (Channel of Sicily): Bulletin Volcanologique, v. 33, p. 828–839.

Walker, G.P.L., 1973, Lengths of lava flows: Royal Society of London Philosophical Transactions, ser. A, v. 274, p. 107–118.

Walker, G. W., and Swanson, D. A., 1968, Laminar flowage in a Pliocene soda rhyolite ash-flow tuff, Lake and Harney counties, Oregon: U.S. Geological Survey Professional Paper 600 B, p. 37–47.

Wolff, J. A., and Wright, J. V., 1981, Rheomorphism of welded tuffs: Journal of Volcanology and Geothermal Research, v. 10, p. 13–34.

MANUSCRIPT ACCEPTED BY THE SOCIETY MAY 5, 1986

Physical features of rhyolite lava flows in the Snake River Plain volcanic province, southwestern Idaho

Bill Bonnichsen
Idaho Geological Survey,
University of Idaho,
Moscow, Idaho 83843

Daniel F. Kauffman
Department of Geology,
University of Idaho,
Moscow, Idaho 83843

ABSTRACT

Large, Miocene-age rhyolite lava flows occur in the Bruneau-Jarbidge area of the central Snake River Plain and in the adjoining Jacks Creek area. The flows typically are 100–150 m thick and have volumes ranging from 10 to 200 km^3. The flow interiors consist of thick central zones of massive devitrified rhyolite overlying zones of basal vitrophyre. These massive central zones are capped by structurally complex upper zones with both glassy and devitrified rhyolite. The upper zones contain gas cavities of varying dimensions and abundance, including swarms of cavities, each a meter or more across. Sheeting joints, in some places accompanied by pencil and dimple joints, are abundant in the upper zones, in the top part of the central zones, and in the marginal parts of the flows. Flow margins consist of bulbous lobes of massive rhyolite separated by steeply to chaotically jointed zones. Few flow margins are less than 25 m thick. The basal and upper zones and the marginal parts of the flows contain abundant breccia formed by en masse flowage and explosive steam release.

All of the rhyolite flows are believed to have erupted from fissures. Most flowed onto preexisting soils and other sedimentary materials. Small amounts of air-fall ash occur beneath a few flows near their eruptive fissures, but these deposits are thinner and less widespread than the fallout ash blankets that are beneath many of the southwestern Idaho welded-tuff units. A combination of high effusion rates, high temperatures, and large volumes probably imparted sufficiently low bulk viscosities to the lavas to allow them to flow away from their eruptive fissures to form sheets instead of steep-sided domes.

Several large-volume, high-temperature, densely welded, ash-flow-tuff sheets occur in southwestern Idaho. These pyroclastic flows may have coalesced into pools of silicate liquid capable of en masse flowage after emplacement. They are similar in appearance to the rhyolite lava flows in that region. However, a combination of physical characteristics can be used to distinguish the two types of flows. Good indicators that a given rhyolite sheet may be a lava flow—rather than a unit emplaced as ash hot enough to form a liquid pool and flow—are the presence of blunt flow margins, abundant basal and marginal flow breccias, pervasive flow layering, laterally persistent zones of mismatched vertical shrinkage joints, complex contacts between basal vitrophyres and overlying zones of devitrified rhyolite, and abundant zones with pencil jointing, combined with the absence of lithic fragments, pumice fragments, bubble-wall shards, extensive phenocryst breakage, internal subhorizontal ash-emplacement layering, and subparallel flow marks.

INTRODUCTION

Large-volume (up to 200 km^3), Miocene-age, rhyolite lava flows in the western part of the Snake River Plain volcanic province apparently erupted at high discharge rates and at comparatively high temperatures (perhaps 900°–1000°C) for rhyolite. These flows share many physical, chemical, and petrographic characteristics that distinguish them from rhyolite flows and domes outside the Snake River Plain volcanic province and from the voluminous, densely welded, ash-flow sheets in the western part of the province.

In this paper we describe and document by photographs the nature of the small- to large-scale physical features in these lava flows, and present a descriptive model relating these features to their positions within the flows and their sequence of development. We discuss the similarities and dissimilarities of the rhyolite lava flows to the associated high-temperature, welded-tuff units and suggest how to distinguish the two types. In most cases, lava flows may be distinguished from ash flows by the structural features they contain, but in extreme cases where total remobilization of a tuff occurs, ash flows may develop many of the lava flow structures we have described. The information in this paper was collected during our studies in the Bruneau-Jarbidge area (Bonnichsen, 1981, 1982a, 1982b, 1982c; Bonnichsen and Jenks, in press; Jenks and Bonnichsen, in press) and in the Jacks Creek area (Dan Kauffman, M.S. thesis in prep.; Kauffman and Bonnichsen, in press).

GEOLOGIC SETTING

The Snake River Plain geologic province is a large-scale, tectonic, volcanic, and physiographic feature of Neogene age which stretches northeastward across southern Idaho from the region where Idaho, Nevada, and Oregon meet, to Yellowstone National Park in northwestern Wyoming (Fig. 1). The province is oriented nearly perpendicular to the western margin of North America, and forms the boundary zone between two geologic terranes with differing Cenozoic histories. In the Basin and Range province to the south, east-west crustal extension has occurred since the Miocene. To the north, the northern Rocky Mountain–Idaho batholith crustal block has not undergone comparable crustal shortening.

The Snake River Plain volcanic province was formed during an interval of time-transgressive bimodal volcanism which advanced northeastward along a southwest-northeast axis. Included within the Snake River Plain volcanic province are the Snake River Plain (a fairly narrow, arcuate lowland), the Owyhee Plateau (a dissected region of somewhat higher elevations), and the margins of the surrounding mountainous areas. The main axis of this volcanism coincides with the eastern Snake River Plain. In the west, however, the main axis of volcanism did not follow the physiographic Snake River Plain; instead, it extends southwestward under the Owyhee Plateau to near where Idaho, Nevada, and Oregon meet. The western Snake River Plain coincides with a complex graben oriented almost perpendicular to the main axis of volcanism. Our studies have been conducted in the central part of the Snake River Plain where the western Snake River Plain graben joins the main southwest-northeast axis of volcanism.

Extensive silicic volcanism accompanied the formation of both the main axis of volcanism and the western Snake River Plain graben. Both ash-flow tuff sheets and rhyolite lava flows were erupted. In the region where the western Snake River Plain graben joins the main axis of volcanism, the Bruneau-Jarbidge eruptive center was formed when the Cougar Point Tuff units were erupted (Bonnichsen, 1982a; Bonnichsen and Citron, 1982). The resulting basin was partially filled by the rhyolite lava flows that are the focus of this paper. Additional rhyolite lava flows were erupted at the same time in the Jacks Creek area, a volcanic center located along the southwestern margin of the western Snake River Plain immediately northwest of the Bruneau-Jarbidge area (Fig. 1). Similar rhyolite lava flows have been erupted to the southwest in the Owyhee-Humboldt eruptive center (Bonnichsen, 1985), to the northwest in the western Snake River Plain graben, and to the northeast in the Snake River Plain at other times.

RHYOLITE LAVA FLOWS

In this section we note the locations, stratigraphic sequences, ages, and sizes of the rhyolite lava flows in the Bruneau-Jarbidge and Jacks Creek areas and summarize the same information for a few other southwestern Idaho rhyolite lava flows. Silica contents and petrographic data for all of these lava flows are given in Tables 1 and 2 so that the units can easily be compared to one another, as well as to some of the densely welded ash-flow sheets in the region.

Ekren and others (1984, p. 49–53) interpreted several of the rhyolite lava flows in the Bruneau-Jarbidge area as densely welded ash-flow sheets which flowed like lava after emplacement. They suggested that these units are an extension of their Little Jacks Tuff. Furthermore, they (Ekren and others, 1981, 1982, p. 229–231, 1984, p. 42–49) interpreted the rhyolite lava flows in the Jacks Creek area as ash-flow tuff sheets. On the basis of the various indicators we have used to distinguish rhyolite lava flows from ash-flow tuff units, stratigraphic relations, and our geologic mapping (Bonnichsen and Jenks, in press; Jenks and Bonnichsen, in press; Kauffman and Bonnichsen, in press; Kauffman, M.S. thesis in prep.), we do not agree with their interpretation for any of these units.

Bruneau-Jarbidge Area Flows

The rhyolite lava flows in the Bruneau-Jarbidge eruptive center lie beneath numerous basalt flows, shield volcanoes, and accumulations of lacustrine, fluvatile, and fanglomeratic sediments, and above the Cougar Point Tuff, which previously was erupted from the same volcanic center (Bonnichsen, 1982a, 1982b; Bonnichsen and Citron, 1982). The rhyolite lava flows

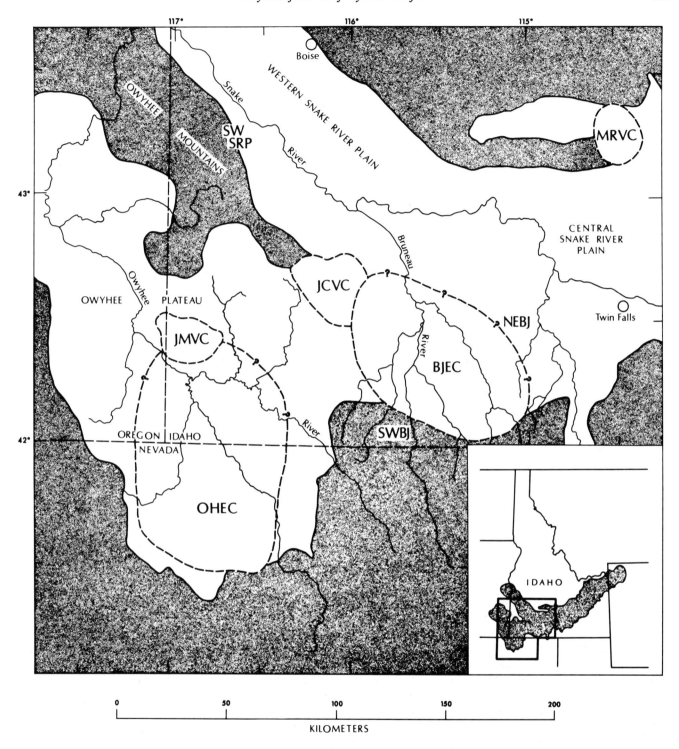

Figure 1. Map of southwestern Idaho and vicinity showing centers from which rhyolitic rocks were erupted in the western part of the Snake River Plain volcanic province. The inset map shows the location of the larger map in relation to the combined Snake River Plain, Owyhee Plateau, and Yellowstone Park regions. Abbreviations used are: BJEC, Bruneau-Jarbidge eruptive center; JCVC, Jacks Creek volcanic center; JMVC, Juniper Mountain volcanic center; MRVC, Magic Reservoir volcanic center; NEBJ, area northeast of Bruneau-Jarbidge eruptive center; OHEC, Owyhee-Humboldt eruptive center; SWBJ, area southwest of Bruneau-Jarbidge eruptive center; SWSRP, southwestern margin of western Snake River Plain.

TABLE 1. CHARACTERISTICS OF RHYOLITE LAVA FLOWS IN THE BRUNEAU-JARBIDGE ERUPTIVE CENTER, SOUTHWESTERN IDAHO

Name of Unit	Phenocryst Minerals	SiO_2	N	Maximum Thickness (m)	Minimum Volume (km^3)
Three Creek Rhyolite	plag qtz aug pig ox	73.7	1	---	10
rhyolite, Juniper-Clover area	plag qtz aug pig ox	73.1	4	---	very large
Dorsey Creek Rhyolite	plag qtz aug pig ox	72.7	6	>200	75
Poison Creek Rhyolite	plag qtz aug pig ox	70.7	2	about 100	---
Sheep Creek Rhyolite	plag qtz aug pig ox	70.4	6	about 250	200
Bruneau Jasper Rhyolite	plag san qtz aug pig ox	73.9	3	about 150	---
lower rhyolite, Louse Creek	plag aug pig ox	70.1	2	---	---
Long Draw Rhyolite	plag aug pig ox	71.0	3	about 100	10
Cedar Tree Rhyolite	plag san qtz aug pig ox	72.5	2	>100	---
Marys Creek Rhyolite	plag san qtz aug pig ox	72.0	2	---	large
Indian Batt Rhyolite	plag aug pig ox	70.1	3	>100	12
Triguero Homestead Rhyolite	plag aug pig ox	71.1	4	about 100	10

Average SiO_2 = 71.8

Notes:
1. The units are listed in order of increasing age. The relative ages of the Three Creek Rhyolite, the rhyolite of the Juniper-Clover area, and the Marys Creek Rhyolite are not known. The Long Draw Rhyolite and the lower rhyolite at Louse Creek may be correlative.
2. The phenocryst mineral abbreviations are: plag = plagioclase, san = sanidine, qtz = quartz, aug = augite, pig = pigeonite, and ox = opaque oxides.
3. The SiO_2 values are weight % SiO_2 normalized to a sum of 100 % for the 10 major constituent oxides (SiO_2, Al_2O_3, TiO_2, Fe_2O_3, MnO, CaO, MgO, K_2O, Na_2O, and P_2O_5); all iron expressed as Fe_2O_3.
4. N is the number of analyzed rocks that were averaged for the SiO_2 values.
5. References: Bonnichsen, 1982b, 1982c; Bonnichsen and Jenks, 1986; Jenks and Bonnichsen, 1986; Hart and Aronson, 1983; Bonnichsen, unpublished data.

are listed in stratigraphic order in Table 1, and their locations are shown in Figure 2. Further details on their stratigraphic succession and map distribution are available in Bonnichsen (1981, 1982a, 1982b, 1982c), Bonnichsen and Jenks (in press), and Jenks and Bonnichsen (in press). Most of the rhyolite lava flows in the Bruneau-Jarbidge area are exposed continuously in canyon walls or between the canyons, and all of these flows are essentially flat lying and undeformed, except for slight normal-fault displacements. We have noted below the distances that individual flows have been followed, in order to present an impression of their dimensions.

The Triguero Homestead Rhyolite (TH) is exposed for more than 12 km in Bruneau River canyon (Fig. 3) and extends 8 km westward to Sheep Creek. The Indian Batt Rhyolite (IB) is exposed for 16 km in Bruneau River canyon (Fig. 3) and for 19 km in Sheep Creek canyon. The Marys Creek Rhyolite (MC) has been traced for 25 km along the base of the Grasmere escarpment and extends 5 km westward in a graben. The Cedar Tree Rhyolite (CT) is exposed for more than 6 km in the bottom of Bruneau River canyon. The Long Draw Rhyolite (LD) is exposed for more than 6 km in Bruneau River canyon and for about 4 km in Jarbidge River canyon. To the west, the lower rhyolite at Louse Creek (LC) is exposed in Sheep Creek canyon for 8 km. Similarities in chemistry, phenocrysts, and magnetic polarity of the Long Draw Rhyolite and the lower rhyolite at Louse Creek would permit the two to be different parts of the same flow. The Bruneau Jasper Rhyolite (BJ) is exposed for 8 km in the bottom of Bruneau River canyon. The Sheep Creek Rhyolite (SC) is exposed in Bruneau River canyon (Fig. 4) for 30 km, in Sheep Creek canyon for 25 km, and northwestward as far as Big Jacks Creek. This flow extends 42 km from southeast to northwest, and exceeds 200 m in thickness throughout much of its extent; it is the largest in the region, and has an estimated minimum volume of about 200 cu km. The Poison Creek Rhyolite (PC) is exposed for

TABLE 2. CHARACTERISTICS OF RHYOLITE LAVA FLOWS IN THE JACKS CREEK AREA AND AT OTHER LOCALITIES IN SOUTHWESTERN IDAHO

Name of Unit	Area	Phenocryst Minerals	SiO_2	N	Maximum Thickness (m)	Flow area (km^2)	Ref.
Horse Basin Rhyolite	JCVC	plag qtz aug pig ox	71.5	2	150	66	1
Perjue Canyon Rhyolite	JCVC	plag aug pig olv ox	74.4	9	150	150	1
Tigert Springs Rhyolite	JCVC	plag san aug pig olv ox	74.0	9	110	450	1
O X Prong Rhyolite	JCVC	plag san pig ox	73.5	8	150	>90	1
Rattlesnake Creek Rhyolite	JCVC	plag san aug pig olv ox	73.4	2	100	90	1
Jump Creek Rhyolite	SWSRP	plag san qtz cpx hyp olv ox	71.1	1	---	474	2
Circle Creek Rhyolite	OHEC	plag san qtz aug olv ox	74.0	1	---	>100	3
Johnstons Camp Rhyolite	SWBJ	plag san aug pig ox	73.5	2	60	20	4
upper rhyolite, Crows Nest area	NEBJ	plag aug pig hyp ox	70.9	3	---	---	5
rhyolite, Balanced Rock area	NEBJ	plag aug pig hyp ox	69.6	3	---	---	5
			Average SiO_2 = 72.6				

Notes:
1. The units in the Jacks Creek area are listed in order of increasing age.
2. The area abbreviations are: JCVC = Jacks Creek volcanic center, SWSRP = southwestern margin of western Snake River Plain, OHEC = Owyhee-Humboldt eruptive center, SWBJ = area southwest of Bruneau-Jarbidge eruptive center, NEBJ = area northeast of Bruneau-Jarbidge eruptive center.
3. The phenocryst mineral abbreviations are: plag = plagioclase, san = sanidine, qtz = quartz, aug = augite, pig = pigeonite, cpx = clinopyroxene, hyp = hypersthene, olv = olivine, and ox = opaque oxides.
4. The SiO_2 values are weight % SiO_2 normalized to a sum of 100 % for the 10 major constituent oxides (SiO_2, Al_2O_3, TiO_2, Fe_2O_3, MnO, CaO, MgO, K_2O, Na_2O, and P_2O_5); all iron expressed as Fe_2O_3.
5. N is the number of analyzed rocks that were averaged for the SiO_2 values.
6. References: 1 = Kauffman and Bonnichsen, 1986; Kauffman, unpublished data
2 = Ekren and others, 1984; Kittleman and others, 1965
3 = Coats, 1968
4 = Bernt and Bonnichsen, 1982; Bonnichsen and others, 1986, Bonnichsen, unpublished data
5 = Bonnichsen, 1982b, 1982c; Bonnichsen, unpublished data

4 km in Jarbidge River canyon. The Dorsey Creek Rhyolite (DC) is exposed for 40 km in Jarbidge River canyon and for 12 km in Bruneau River canyon. It has a minimum volume of 75 cu km, but it may be much larger, because the Three Creek Rhyolite and the rhyolite of the Juniper-Clover area are similar enough in chemistry, phenocrysts, and paleomagnetic polarity to permit either or both to be eastward extensions. The rhyolite of the Juniper-Clover area (JC) has not been eroded deeply enough to tell if one or more flows are present, but if it is one flow, its volume would be comparable to that of the Sheep Creek and Dorsey Creek flows. The Three Creek Rhyolite (TC) underlies an area measuring about 7 by 17 km (Citron, 1976).

The following age information for the rhyolite lava flows brackets their eruptions within the general time limits of 11.3 to 8.0 Ma. These lava flows are younger than the Cougar Point Tuff because the oldest lava flow (Triguero Homestead Rhyolite) lies on the uppermost Cougar Point Tuff unit. Bonnichsen and Citron (1982) reported a whole-rock K-Ar age of 11.3 ±2.0 Ma for one of the lower Cougar Point Tuff units. We believe that all of the Bruneau-Jarbidge rhyolite lava flows are younger than the rhyolite of the Grasmere escarpment, which forms the western margin of the Bruneau-Jarbidge eruptive center (Bonnichsen, 1982a). Hart and Aronson (1983) reported a whole-rock K-Ar age of 11.22 ±0.52 Ma for that unit. They also reported whole-rock K-Ar ages of 9.88 ±0.46 Ma for the Sheep Creek Rhyolite and 8.00 ±0.38 Ma and 8.22 ±0.38 Ma for the Dorsey Creek Rhyolite. These Dorsey Creek dates set the approximate upper time limit for the silicic-flow eruptions in the Bruneau-Jarbidge area.

Jacks Creek Area Flows

Five lava flows constitute the rhyolite of the Jacks Creek area. In decreasing age, these are the Rattlesnake Creek Rhyolite, the O X Prong Rhyolite, the Tigert Springs Rhyolite, the Perjue

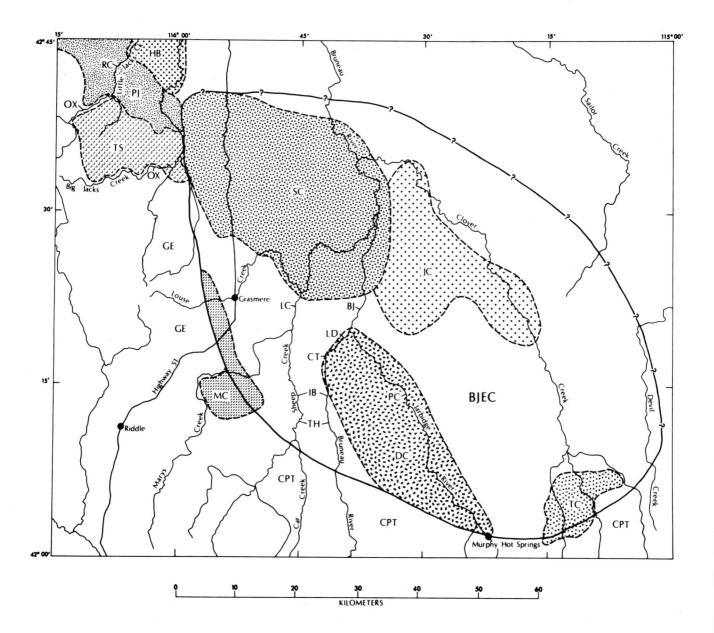

Figure 2. Map showing approximate locations of the rhyolite lava flows and other rhyolitic units in the Bruneau-Jarbidge and Jacks Creek areas. Heavy solid line indicates approximate margin of Bruneau-Jarbidge eruptive center (BJEC). Volcanic units are either outlined by dashed lines and patterned, or located by symbols and leaders where restricted to river canyons. Abbreviations for the individual rhyolite lava flows and other volcanic units are: BJ, Bruneau Jasper Rhyolite; CPT, Cougar Point Tuff; CT, Cedar Tree Rhyolite; DC, Dorsey Creek Rhyolite; GE, rhyolite of Grasmere escarpment; HB, Horse Basin Rhyolite; IB, Indian Batt Rhyolite; JC, rhyolite of Juniper-Clover area; LC, lower rhyolite at Louse Creek; LD, Long Draw Rhyolite; MC, Marys Creek Rhyolite; OX, O X Prong Rhyolite; PC, Poison Creek Rhyolite; PJ, Perjue Canyon Rhyolite; RC, Rattlesnake Creek Rhyolite; SC, Sheep Creek Rhyolite; TC, Three Creek Rhyolite; TH, Triguero Homestead Rhyolite; TS, Tigert Springs Rhyolite.

Figure 3. View toward the north in the Triguero Homestead area of Bruneau River canyon at the Indian Batt Rhyolite (IB) overlying the Triguero Homestead Rhyolite (TH). The rim unit is the Black Rock Basalt (BR). An unnamed lower basalt unit (LB) lies just above the Indian Batt Rhyolite. The white zone at the right below the Triguero Homestead Rhyolite is a thick accumulation of volcanic ash.

Figure 4. View toward the north in Bruneau River canyon near the mouth of Stiff Tree Draw. The central zone of the Sheep Creek Rhyolite is exposed in the cliffs, and the upper zone corresponds to the overlying slope. Basalt flows are exposed at the canyon rim. The central zone of the rhyolite flow is about 200 m thick, and the upper zone is 30 to 50 m thick.

Canyon Rhyolite, and the Horse Basin Rhyolite. Their locations are shown in Figure 2 and further information is available in Kauffman and Bonnichsen (in press) and Kauffman (M.S. thesis in prep.). The rhyolite of the Jacks Creek area was interpreted by Ekren and others (1981, 1982, and 1984) as part of a regionally extensive, multiple-flow, map unit that they called the tuff of Little Jacks Creek and then the Little Jacks Tuff. Intercalated between the O X Prong and Tigert Springs lava flows is the rhyolite of Grasmere escarpment, which is probably a large welded-tuff sheet (Bonnichsen, 1982a). This unit may be equivalent to Cougar Point Tuff unit VII (Bonnichsen and Citron, 1982), as well as constituting a major part of the Little Jacks Creek tuff of Ekren and others (1981).

All of the Jacks Creek area rhyolite lava flows are fresh and undeformed; they dip 3 to 6 degrees northeastward. The upper three units are clearly lava flows, but the lower two are not well enough exposed for complete identification. Because these two lower units contain some indications that suggest they are lava flows, we have included them in our discussion. The four older Jacks Creek area rhyolite units are quite similar in their physical, chemical, and petrographic characteristics (Table 2). The youngest flow, the Horse Basin Rhyolite, has a higher percentage of phenocrysts, a lower silica content, and more breccia than the older units.

The Rattlesnake Creek Rhyolite (RC) extends for 15 km in the bottom of Little Jacks Creek canyon. The O X Prong Rhyolite (OX) is exposed at four separate locations in the Jacks Creek area canyons, including nearly the full length of Little Jacks Creek canyon and most of the length of Big Jacks Creek canyon. The Tigert Springs Rhyolite (TS) is exposed for much of the lengths of these canyons and in the area between. The Perjue Canyon Rhyolite (PJ) covers an extensive area between Big and Little Jacks creeks. The northeastern part of the flow is down-faulted, so the unit's full extent is unknown. The Horse Basin Rhyolite (HB) is exposed for 9 km between Big Jacks Creek and Little Jacks Creek, where its original southwestern margin is exposed along an escarpment. This flow evidently was erupted from a buried, northwest-trending fissure near its southwestern margin. This probable venting area is partially exposed in Big Jacks Creek canyon.

Additional Rhyolite Lava Flows

In addition to the flows in the Bruneau-Jarbidge and Jacks Creek areas, other Miocene-age rhyolite lava flows are present in or near southwestern Idaho. Five of them are noted below and their petrographic constituents and silica contents are included in Table 2. All five are similar in their physical and petrologic features to the flows in the Bruneau-Jarbidge and Jacks Creek areas. Rhyolite lava flows and large domes also occur in the central and eastern parts of the Snake River Plain volcanic province, including those described by Leeman (1982) and Struhsacker and others (1982) in the Magic Reservoir volcanic center (Fig. 1), by Spear and King (1982) at Big Southern Butte, and by Christiansen (1982) in the Island Park area.

Several rhyolite lava flows are exposed immediately northeast of the Bruneau-Jarbidge area, in a 30 by 50 km zone (general area in Fig. 1 marked by symbol NEBJ). Data for two units there, the upper rhyolite of the Crows Nest area and the rhyolite of the Balanced Rock area, are included in Table 2.

The Johnstons Camp Rhyolite is exposed a few kilometers southwest of the Bruneau-Jarbidge area (general area in Fig. 1 marked by symbol SWBJ), and probably contains about 1 km^3 of rhyolite. This unit is older than the Cougar Point Tuff (Bernt and Bonnichsen, 1982; Bonnichsen and others, in press) and therefore is older than the Bruneau-Jarbidge area flows.

Farther southwest, in northern Nevada, Coats (1968) has described the Circle Creek Rhyolite, which is located in the eastern part of the Owyhee-Humboldt eruptive center (Fig. 1). The Circle Creek Rhyolite has a sanidine K-Ar age of 11.9 ±0.5 Ma (recalculated for new decay constants by method of Dalrymple, 1979).

The Jump Creek Rhyolite (Kittleman and others, 1965; Ekren and others, 1984) is located along the southwestern side of the western Snake River Plain (Fig. 1). Armstrong and others (1980) reported a K-Ar sanidine age of 11.1 ±0.2 Ma for this unit, and Ekren and others (1984), on the basis of the compositions of coexisting magnetite and ilmenite, reported an approximate eruption temperature of 1100°C for the unit. The Jump Creek Rhyolite probably contains several tens of km^3 of rhyolite.

Petrography

The rhyolite lava flows in the Bruneau-Jarbidge and Jacks Creek areas that we have investigated, and the other flows in southwestern Idaho, have similar phenocryst assemblages and groundmass characteristics. In this section we note phenocryst and textural features that may help indicate the origin and emplacement processes of the rhyolite lava flows, and to which we later refer as evidence for their non-explosive and high-temperature emplacement. Information on the phenocrysts in particular flows is summarized in Tables 1 and 2, and additional petrographic information is available in Bonnichsen (1982b) and Kauffman (M.S. thesis in prep.).

Plagioclase, clinopyroxene, and opaque oxides are the principal phenocryst minerals. These minerals occur in all of the flows and may be accompanied by quartz or sanidine or both. Fayalite and ferrohypersthene have been found in a few flows. Plagioclase-pyroxene-opaque-oxide cumulophyric aggregates, with metamorphic or plutonic igneous textures, are present within most, if not all, of the rhyolite flows. Many have been interpreted as source-rock fragments that were carried up by the magmas (Bonnichsen, 1982b).

Plagioclase is the most abundant phenocryst mineral and is mainly oligoclase and andesine. The most notable characteristic of plagioclase is that grains with different textural forms typically occur together in nearly all thin sections. Single plagioclase crystals are common; these generally are subhedral tablets or laths that probably are phenocrysts that crystallized during magma ascent. Some grains have been extensively embayed and others contain glass inclusions. Most plagioclase crystals in the cumulophyric aggregates have anhedral, internally complex, forms. Anhedral plagioclase crystals with patchy or irregular zoning and grains with anhedral pyroxene and opaque-oxide inclusions may be refractory material that was not entirely melted during magma formation.

The pyroxenes in most flows are a mixture of augite and pigeonite grains that are equant to prismatic single crystals and equant to irregular crystals within the cumulophyric aggregates. Some pyroxene grains are wholly enclosed within plagioclase, and in some aggregates the pyroxene grains enclose rounded opaque oxide grains. Ferrohypersthene and fayalite have been found only in flows from outside of the Bruneau-Jarbidge eruptive center.

Opaque oxides occur in all of the flows. In polished sections from some of the Bruneau-Jarbidge area flows, both magnetite and ilmenite are present, with magnetite generally more abundant. Both magnetite and ilmenite vary from equant to irregular in shape and both occur as single crystals and within the cumulophyric aggregates.

Quartz phenocrysts are present in some flows as small dipyramidal grains that vary from euhedral to rounded and may be embayed. Sanidine phenocrysts generally occur as single crystals, although in a few flows the mineral partially replaces, or is attached to, plagioclase grains. It is rare for either quartz or sanidine to be part of a cumulophyric aggregate.

The groundmass varies from glassy to devitrified. Where glassy, it is commonly gray and contains abundant crystallites, or is brown with few, if any, crystallites. Most vitrophyres show flow layering, which may consist of alternate gray, crystallite-rich, and brown, crystallite-poor, layers. Samples with only brown glass have little or no layering, whereas layering is conspicuous in many gray-glass samples.

The flow layers are deformed around phenocrysts and some layers are folded into patterns which suggest that the phenocrysts were rotated. In some rocks the flow layers were broken by microfaults and microboudinage. Occasional phenocrysts were broken during flowage in some of these same rocks. These textures have been found mainly in the most distal parts of the flows where there is other evidence of flowage and concomitant breakage. Presumably, these textures all resulted from high lava viscosities.

Instead of flow layers, a few vitrophyre samples have globule textures where one color of glass encloses deformed particles of another color of glass. These textures may record pyroclastic activity, although it is not clear if they formed when the flows were erupted or as they traveled across the surface.

In partly devitrified rhyolite, it is common for microspherulites to be suspended in the glassy matrix. These may be accompanied by devitrified zones at the edges of the plagioclase grains and other phenocrysts. Less commonly, devitrification has occurred along selected flow layers in glassy material or in patchy zones. Where devitrification has been complete, the mafic minerals are oxidized.

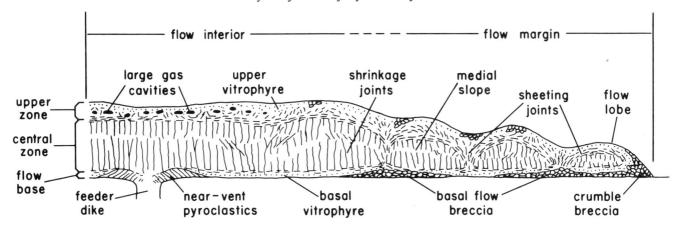

Figure 5. Schematic diagram illustrating the spatial relationships among the zones and physical features that are typical of southwestern Idaho rhyolite lava flows. This idealized longitudinal section combines features found in many of the flows. The vertical scale and the size of some features have been exaggerated.

Chemical Composition

The rhyolite lava flows vary in chemical composition but share characteristics suggesting similar modes of origin. The flows are calc-alkaline and most contain 70-74% SiO_2 (Tables 1 and 2), have more K_2O than Na_2O, and show variable coherent enrichments in femic constituents (Fe, Ti, P, Ca, Mn, and Mg) that are antithetic to the SiO_2 and K_2O behavior (Bonnichsen, 1982b, 1982c; Kauffman and Bonnichsen, in press). The least siliceous flows, those with less than 72% SiO_2, have 4% or more iron oxides, with some exceeding 5%. For example, the six Sheep Creek Rhyolite samples noted in Table 1 average 5.35% total Fe calculated as Fe_2O_3, which is more iron than in most rhyolites.

Inspection of about a hundred whole-rock analyses (Bonnichsen, 1982b, 1982c; Bernt and Bonnichsen, 1982; Hart and Aronson, 1983; Ekren and others, 1984; Bonnichsen, unpublished data; Kauffman and Bonnichsen, in press; Kauffman, unpublished data) shows that the lava flows are more restricted in composition individually than as a group.

PHYSICAL NATURE OF RHYOLITE LAVA FLOWS

In this section we describe the physical features in the southwestern Idaho rhyolite lava flows and comment on their possible modes of origin. We first discuss the overall configuration and character of the flows, and then describe their constituent parts and smaller features. Our discussion is accompanied by photographs of specific features and by Figure 5, which is a generalized longitudinal section from vent area to distal margin in an idealized flow.

Overall Aspects

A rhyolite lava flow interior consists of three zones: a thick central zone of massive devitrified rhyolite which overlies a complex basal zone and which is capped by a structurally complicated upper zone. The basal zones contain massive vitrophyre or breccia or a combination of both, and the upper zones generally contain both glassy and devitrified portions and consist of massive, sheeted, folded, flow-layered, and brecciated rhyolite. All of the zones, but especially the upper zones, contain gas cavities of widely varying dimensions and abundance. Flow margins typically consist of bulbous lobes of massive or sheeted rhyolite overlying basal breccia layers and separated by chaotic-appearing zones of steeply jointed or flow-layered rhyolite.

Thicknesses of the interior portions of the flows typically are 50 m or more and range up to 250 m. The flows are thinner near their margins, but few are less than 25 m thick. Many flows are 100 to 150 m thick throughout their exposures (Tables 1 and 2).

Central Zones

The rhyolite in the massive central zones of the flows is generally uniform, dense, and devitrified, although gas cavities, folded flow layering, breccia zones, and even vitrophyres are sporadically present. The central zones contain well-developed, vertical or locally inclined shrinkage joints that divide the flows into elongate columnar masses a few to several meters across (Figs. 4, 6, and 7). As illustrated in Figure 6, shrinkage joints characterize devitrified rhyolite and are distinct from the columnar joints that are sporadically present in the glassy portions of the flows. Subhorizontal sheeting joints are also present in the central zones and are more conspicuous near the tops and bases of those zones.

In most rhyolite flows the central zones have eroded into two cliffs that are separated by a medial interval of devitrified rhyolite where late-stage shrinkage joints are more numerous (Figs. 3, 4, 5, and 7). This medial interval erodes to a steep slope because the numerous shrinkage joints are inclined at various angles. Although the double cliffs are most conspicuous in the

Figure 6. The lower part of the Triguero Homestead Rhyolite in Bruneau River canyon. The basal vitrophyre zone is cut by closely spaced vertical columnar joints, and the overlying lower portion of the devitrified central zone contains subhorizontal sheeting joints and widely spaced vertical shrinkage fractures. Note that the two vertical joint sets cut different parts of the flow.

Figure 7. View toward the northwest in Jarbidge River canyon 7 km upstream from its mouth. The Dorsey Creek Rhyolite (DC) overlies the Indian Springs Basalt (IS). The cliff with the prominent vertical joints is the central zone of the Dorsey Creek Rhyolite, and the overlying snow-covered slope is eroded in the upper zone of this rhyolite flow. On the distant canyon rim are basalt flows.

thickest portions of the flows, they also occur in thinner parts and even in some of the marginal flow lobes (Fig. 8). Their distribution suggests that the shrinkage joints propagated toward the flow interiors from both the upper and lower surfaces. These joints generally do not continue through the medial slopes (Figs. 4, 5, and 7), implying that the medial intervals are the most slowly cooled parts of the flows.

Basal Zones

The basal zones of the flow interiors are made of massive, flow-layered vitrophyre (Fig. 9), breccia (Figs. 5, 10, and 11), or a combination of vitrophyre and breccia. The basal zones are generally 2 to 5 m thick but may exceed 10 m.

Beneath the flow margins, the basal zones are mostly breccia consisting of roughly equant vitrophyre blocks in a matrix of finer fragments and ashy material (Fig. 10). This matrix may either be made of loose material or may be thoroughly fused to form a competent rock. Most basal breccias probably originated as crumble breccias at the steep fronts of slowly moving masses of rhyolitic lava, which were overridden as the lava advanced (compare Figs. 10 and 12). In a few places, the basal breccias consist of angular devitrified fragments (Fig. 11), indicating that parts of the flows were devitrified before incorporation into the basal breccias.

In the interiors of the flows, the basal zones mostly consist of uninterrupted vitrophyre layers. These contain local, tightly folded, flow layering (Figs. 9 and 13), which has subhorizontal axial planes and sparse, greatly stretched, gas cavities. At some exposures the flow-layered rhyolite may be a mixture of glassy

Figure 8. A flow lobe in the Dorsey Creek Rhyolite near the mouth of Poison Creek. Note the double cliff, the vertical shrinkage fractures, and the sheeting joints that wrap around the upper right-hand side of the lobe. The poorly exposed unit below the flow lobe is the Poison Creek Rhyolite, and the canyon rim is basalt.

and devitrified rhyolite (Fig. 13). In parts of some flows, layers of both massive vitrophyre and flow breccia occur together at the base (Fig. 5). The vitrophyre layer is usually above the basal breccia in these situations.

Lenses and irregular zones of breccia that probably formed in place are also present within otherwise massive basal vitrophyre layers (Fig. 14). Figure 14 shows how tightly the fragments are packed in comparison with the much looser packing in an overridden, crumble type of breccia (Fig. 10). The tight packing

Figure 9. The basal vitrophyre of the Indian Batt Rhyolite in Bruneau River canyon. The flow lies directly on a baked soil horizon (below hammer). Note the large spherulites in the upper part of the view and the subhorizontal flow layers in the middle.

Figure 10. Flow breccia at the base of the Dorsey Creek Rhyolite in Jarbidge River canyon. This breccia of vitrophyre blocks in an ashy matrix probably formed as the lava advanced over blocks that had crumbled from the flow front.

Figure 11. Basal flow breccia of the Dorsey Creek Rhyolite (above knife) exposed near the flow margin in Jarbidge River canyon. The rhyolite was devitrified before it was fragmented; note angularity of the blocks. The fine-grained sediments beneath the flow (below knife) are shallow-lake deposits that probably were wet when the flow was emplaced and were slightly deformed.

and particle angularity suggest that fragments in the type of breccia shown in Figure 14 have undergone little, if any, transport. Such in-place brecciation probably resulted from a combination of stresses caused by movement in the overlying flow and the expansion of steam from water heated beneath or within the basal part of the flow. In some flows these tightly packed breccia zones at the flow bases grade into red-and-black breccias resembling those in the upper zones that we will describe later.

Devitrification spherulites are common in the basal zones of the flows and are most abundant near the tops of the vitrophyre layers (Fig. 9). Some are as large as a meter in diameter. Many, especially the larger ones, contain angular shrinkage cavities.

The contacts between the basal vitrophyre layers and the overlying devitrified rhyolite are typically sharp and have general subhorizontal attitudes that can be followed for long distances (Fig. 15). In detail, these contacts are slightly to extremely irregular. Numerous exposures of glassy zones extending for several meters into the overlying devitrified material, of interlayered glassy and devitrified rhyolite (Fig. 13), and of a mixture of glass and devitrified rhyolite enclosing spherulites and breccia fragments have been noted.

Columnar joints occur in the glassy portions of some rhyolite flows. The Triguero Homestead Rhyolite, for example, has remarkably well-developed columnar joints at its base (Fig. 6). Column attitudes include upright, inclined, horizontal, and fan-shaped; these arrangements probably reflect cooling next to topo-

Figure 12. Equant vitrophyre blocks in a rhyolitic ash matrix at the northern margin of the Triguero Homestead Rhyolite in Bruneau River canyon. This is interpreted as a crumble breccia.

Figure 14. Irregular lenses and tongues of breccia within the basal vitrophyre of the Long Draw Rhyolite in Bruneau River canyon. The breccia appears to have formed in place from the enclosing massive vitrophyre. The packing of the blocks and configuration of the brecciated zones suggest that they may have resulted from a combination of intense stresses at the flow base caused by movement of the overlying material and by explosive steam release.

Figure 13. Flow layering and tight folds in the basal zone of the Cedar Tree Rhyolite in Bruneau River canyon. A mixture of glassy and devitrified rhyolite forms the alternating layers. The axial planes of the folds are nearly parallel to the flow base, which dips toward the right.

Figure 15. The sharp, but irregular, contact between the basal vitrophyre zone (below) and the bottom of the devitrified central zone (above) in the Indian Batt Rhyolite in Sheep Creek canyon. At the upper left, note the ghost of a spherulite and its irregular shrinkage cavity preserved in the devitrified rhyolite.

Figure 16. The boundary between the central zone (lower cliff) and the upper zone (scattered outcrops on slope) of the Sheep Creek Rhyolite at the mouth of Stiff Tree Draw in Bruneau River canyon. Note the prominent subhorizontal sheeting at the top of the central zone and the subhorizontal layer of giant gas cavities within the upper zone.

Figure 17. Thick vitrophyre layer at the top of the lower rhyolite at Louse Creek in Sheep Creek canyon. Note the blocky fracture pattern and the imperfectly developed columnar joints, and the complete lack of sheeting joints.

graphic irregularities. Columnar joints do not occur in devitrified rhyolite nor do they connect with the vertical shrinkage fractures that segment the central zones into larger and more crudely shaped columns (Fig. 6).

The basal vitrophyre and breccia layers typically rest on variably baked, generally structureless, tan, gray, or reddish-brown silt; intervening volcanic ash layers are rare. For example, Figure 9 shows the basal vitrophyre of the Indian Batt Rhyolite lying directly on baked silt. Much of this sediment may be old soil layers. At places the flows rest on fine-grained, white, clay-rich sediments that probably were deposited in shallow lakes, and which may have been wet when the rhyolite flows erupted. In areas where rhyolite flowed onto sediments that apparently were wet, the upper 1 to 2 m of these materials were deformed (Fig. 11), and baked mud is present in the basal-zone breccia matrix of some flows.

Volcanic ash is absent or sparse at the base of most lava flows. Most exposures show none, and where ash layers are present they typically are only a few centimeters thick. The only comparatively thick deposits are beneath the Triguero Homestead and Horse Basin flows; these probably were deposited near the vent areas.

Upper Zones

The upper zones range from a few meters to perhaps 50 m in thickness and are structurally complex. They contain glassy and devitrified rhyolite, dense to highly vesicular rock, folded, flow-layered, and variably jointed material, several types of breccias, and silica deposits that formed as the flows cooled. In the flow margins the upper zones merge into similarly complex septa between flow lobes, where steeply dipping sheeting joints are common. Because of their greater susceptibility to erosion, the upper zones are not as well exposed as the central zones in canyon walls (Fig. 16).

Generally, the top part of an upper zone is black and rather massive vitrophyre (Fig. 17), but it may be moderately vesicular glass that is oxidized to red or brown. Blocky breakage or crude to well-developed columnar joints appear in the vitrophyre in the upper zones (Fig. 17). The glassy rhyolite at the top of the upper zones commonly displays tightly folded flow-layering with steeply dipping axial planes (Fig. 18).

Red-and-black breccias, so named because they consist of black vitrophyre blocks and smaller fragments enclosed in a reddish matrix of comminuted, but fused, glass or devitrified material, are abundant in the upper zones of the rhyolite lava flows. Two red-and-black-breccia types are recognized: the jostle type and the fumarolic type. These types may be intergradational.

The jostle type of breccia consists of equant, cobble- to boulder-sized blocks of black vitrophyre which are closely packed and set in a matrix of oxidized red vitrophyre (Fig. 19). This type of breccia probably was formed when rapidly cooled glassy rhyolite was broken up and jostled about by movement in the underlying hotter and less viscous lava. Flow layering and folds are preserved in many of the black vitrophyre clasts. We suggest that the formation and oxidation of the matrix resulted from both the jostling of the blocks and explosive steam venting

Figure 18. Folded flow layers in the upper vitrophyre of the Cedar Tree Rhyolite in Bruneau River canyon. The flow layers and the axial planes of the folds dip steeply; this attitude is typical near the top of the rhyolite flows.

Figure 19. A breccia of closely packed equant blocks of black vitrophyre in a matrix of red, oxidized vitrophyre from the upper zone of the Dorsey Creek Rhyolite near Jarbidge River canyon. This is the jostle type of red-and-black breccia; it probably results from movement in the lava, which breaks up the upper zone, and from the escape of high-temperature steam, which oxidizes the breccia matrix. Note the folded flow layers in some of the breccia blocks.

caused by the introduction of water into the flow tops. Areas in some upper zones contain breccia in which the jostling did not entirely disrupt the previously formed flow layers and folds in the glassy rhyolite, so they can be traced through adjoining breccia blocks (Fig. 20).

The fumarolic type of breccia is common in the upper zones of all the flows. In this type the black fragments are angular and isolated (Fig. 21), and the matrix consists of fine-grained, fragmented and oxidized glass particles that have been re-fused without being altered to form a competent rock. The boundaries of these breccia zones generally grade into massive black vitrophyre with a gradual change in the proportions of angular blocks to matrix, until the reddish matrix-type material occurs only in thin (1 mm to 1 cm) veins cutting the massive black vitrophyre. This type of breccia probably formed by the explosive release of steam when water entered the flow tops and became superheated as it followed fractures down into the hot flows.

In the upper zones the ratio of devitrified rock to vitrophyre increases downward. This transition is complex; sharp subhorizontal contacts between glassy and devitrified material rarely extend for more than a few meters. In most exposures the glassy and devitrified rhyolite are interlayered or mixed together, and some spherulites and irregularly shaped devitrification structures are present. The spherulites in the upper zones are like those at the flow bases, but are smaller and not as abundant. In some areas the jostle type of red-and-black breccia grades downward to where both the blocks and matrix have been devitrified, leaving only vestiges of the original fragmental structure. This gradation, along with the absence of deformation of the breccia-clast margins, suggests that the devitrification of the breccias took place after the flows had stopped moving.

Devitrified rhyolite with abundant sheeting joints is common in the upper zones, especially in the lower parts of the zones and in the septa between the flow lobes at the flow margins. These subparallel joints vary widely in their attitudes and may be contorted (Fig. 22). It is common for the dip of the joints to range from near vertical in their uppermost occurrences to subhorizontal at the boundary between the upper and central zones (Fig. 16).

Dikelike, generally near-vertical masses of glassy rhyolite, up to several meters in width, have been noted between walls of devitrified rhyolite in the upper zones of some flows. These evidently are autointrusive, or rootless, dikes that formed when lava was squeezed up into fractures from the flow interior. Some of these dikes have been devitrified and some are partially to completely brecciated. Bulbous protrusions of rhyolite, also probably caused by late-stage lava squeeze-ups, are sporadically present in the upper zones. These resemble the larger flow lobes at the flow margins.

Flow Margins

Most flow margins consist of a series of bulbous flow lobes of massive devitrified rhyolite separated by septa of steeply

Figure 20. A breccia of flow-layered vitrophyre fragments set in a partially altered glassy matrix in the upper zone of the Cedar Tree Rhyolite flow in Bruneau River canyon. Note that not all of the fragments have moved from their original positions, thus preserving part of a fold.

Figure 22. Steeply dipping and slightly folded sheeting joints developed in devitrified rhyolite near the margin of the Marys Creek Rhyolite near Marys Creek canyon.

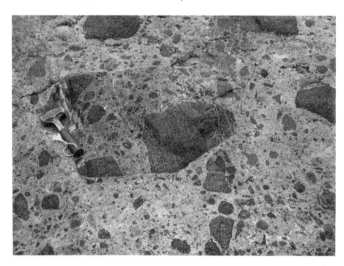

Figure 21. A typical fumarolic type of red-and-black breccia from the upper zone of the Cedar Tree Rhyolite at the mouth of Long Draw in Bruneau River canyon. In this type of breccia, the angular fragments of black glassy rhyolite are separated, and set in a reddish matrix of comminuted and re-fused, oxidized but unaltered, glassy rhyolite. This type of breccia results from the explosive escape of high-temperature steam at the flow tops. Note the large angular block of refragmented breccia beside the hammer.

sheeted, folded, brecciated, and variably vesicular, devitrified rhyolite and vitrophyre (Figs. 5, 8, and 23). The complex combination of these diverse structural forms causes the flow margins to appear chaotic rather than laterally continuous. The marginal zones typically are 0.5 to 2 km wide and merge into the flow interiors as the amount of structurally complex rhyolite between adjacent lobes diminishes (Figs. 3, 4, 5, and 7). Commonly, the

Figure 23. A flow lobe in the Dorsey Creek Rhyolite in Jarbidge River canyon about 6 km upstream from the mouth of the river. Note how the shrinkage fractures dip at various angles, all approximately perpendicular to the margin of the lobe. The lobe is about 150 m high.

marginal zones of the flows range from 50 to 100 m in thickness; it is uncommon for any to be thinner than 25 m. Some units, such as the Sheep Creek, Dorsey Creek, Long Draw, and Horse Basin flows, have numerous and prominent flow lobes. Others, such as the Indian Batt, Three Creek, Perjue Canyon, and Tigert Springs, have fewer and less conspicuous marginal lobes and tend to be thinner at their margins than the flows with conspicuous lobes. A few flow lobes have been found well within the flow interiors (Fig. 23), and may have formed where lava became ponded against a previously emplaced part of a flow.

Flow breccias occur at the bases of the lobes, and crumble breccias are preserved at the margins of some flows. In contrast to the flow interiors, the basal zones of the flow margins contain more overridden crumble breccias than massive vitrophyre. Many meters of crumble breccias are preserved at the edges of some flows. For example, the north end of the Triguero Homestead Rhyolite is a breccia consisting of poorly stratified to nonstratified ash enclosing vitrophyre blocks (Fig. 12). A similar thick breccia accumulation is present at the northern edge of the Horse Basin Rhyolite flow (Ekren and others, 1984, Fig. 26).

The upper and central zones of flow margins have the same variety of structures as occur in the flow interiors. Double cliffs are present in flow lobes (Fig. 8), and their medial slopes may curve downward near the lobe edges, mimicking the shape of their tops. Shrinkage joints at the edges of the lobes usually dip steeply, but in some places they dip shallowly. These joints apparently formed approximately perpendicular to the lobe margins (Fig. 23).

Vent Areas

We believe that the southwestern Idaho rhyolite lava flows erupted from fissures. This conclusion is based on the lack of any obvious rhyolite-vent landforms in the region, rather than on direct observation of fissures or dikes. In a few places, substantial volcanic-ash accumulations beneath the flows suggest that vents are nearby. Elsewhere, in the thick central parts of flow interiors, silicification and dikelike zones of devitrified rhyolite cut the upper vitrophyre zones; these features may mark the locations of buried fissures. The nature of the rhyolite vents is poorly understood. The few observations we have made concerning possible vent areas are noted below.

In the Horse Basin Rhyolite, the portion of the flow in the probable venting area in Big Jacks Creek canyon shows about 4 m of white ash overlain by almost 150 m of breccia, essentially the full thickness of the unit at that locality. The matrix of this breccia is somewhat altered. A few kilometers from this area, the Horse Basin Rhyolite is underlain by only a few centimeters of fine white ash.

Several meters of volcanic ash containing pumice clasts, vitrophyre blocks, and locally slumped bedding are present beneath the Triguero Homestead Rhyolite (Fig. 3) in Bruneau River canyon. These features suggest the eruptive fissure is nearby.

In the interior of the Cedar Tree Rhyolite, abundant ash

Figure 24. The transition zone between sheeted devitrified rhyolite (on the left) and massive vitrophyre (on the right), in the upper zone of the Indian Batt Rhyolite in Sheep Creek canyon. The restriction of the sheeting joints to the devitrified rhyolite suggests that they formed during or after devitrification.

containing vitrophyre blocks is present at the top of the upper zone. This part of the flow appears to have been the topographically high part of the flow after it erupted. Downward, the amount of ash decreases as the amount of vitrophyre increases, until all of the material is vitrophyre. In that same area, silicified, parallel, dikelike ribs of devitrified rhyolite cut the upper vitrophyre of the flow. These dikelike bodies may be over the area from which this flow erupted.

Along the west side of Little Jacks Creek, opposite the mouth of Rattlesnake Creek, the Perjue Canyon Rhyolite flow is about 150 m thick and apparently was ponded in a topographic low. This is the thickest part of the flow and may be over its venting area. This part of the flow is capped by an extra layer of devitrified rhyolite that stands topographically above the rest of the flow, but which grades downward without a devitrification break into the central zone of the flow. Instead of containing the contorted flow layers and sheeting joints that typify most upper zones of flows, this capping zone has horizontal flow layers and lacks sheeting or vertical cooling joints. The layering drapes over the edges of the cap to merge into the underlying central zone of the flow. This cap may have been formed when a late pulse of lava erupted onto the top of the hot flow.

Sheeting Joints and Related Structures

During and after the final stages of flowage and while the rhyolite lavas cooled and devitrified, a variety of joints and related structures were formed. These include sheeting, pencil, and dimple joints, tension gashes, flow marks (parallel streaks on sheeting-joint surfaces), columnar jointing and blocky breakage formed while the rocks were glassy, and shrinkage joints. Shrink-

Figure 25. The development of pencil jointing near the margin of the Sheep Creek Rhyolite in Big Jacks Creek canyon. Here, several somewhat irregular sets of sheeting joints intersect in the devitrified rhyolite.

Figure 26. Dimple joints near the margin of the Sheep Creek Rhyolite in Big Jacks Creek canyon. They formed in devitrified rhyolite at the boundary between a zone of closely spaced sheeting joints (below) and rather massive rhyolite with blocky joints (above).

age and columnar joints were described above; their differences are illustrated in Figure 6. Joints are some of the more conspicuous features of the flows because they affected erosion patterns and because most rock fragments that have broken loose are bounded by joint planes.

Sheeting joints are closely spaced, subparallel joints along which devitrified rhyolite breaks into flaggy sheets or smaller platy fragments ranging from a few millimeters to several centimeters in thickness and from a few centimeters to a meter or more across (Fig. 22). Throughout the central zones of the flows, the sheeting generally is subhorizontal (Fig. 16), but in the upper zones (Fig. 22), in the septa between the flow lobes, and at lobe margins (Figs. 5 and 8), it is moderately to steeply inclined. Where the sheeting joints are very closely spaced, as in the central zones, the rock surfaces are crumbly and difficult to sample.

Sheeting joints are essentially restricted to devitrified rhyolite (Figs. 6, 9, 14, 15, 17, 18, and 19), they rarely are present in devitrified rhyolite and, where present, they are imperfect (Fig. 20). Where sheeting is followed from devitrified into glassy rock, the joints can disappear in less than a meter (Fig. 24). The relation illustrated in Figure 24 has been observed at many locations and suggests that sheeting joints form during, or just following, devitrification. The stresses during and after devitrification, which accompanied the final part of en masse flowage, may have caused the sheeting joints to form, and additional stresses associated with the shrinkage accompanying the devitrification process may have also contributed to their development.

The common parallelism of sheeting joints with previously formed flow layers suggests that earlier flowage imparted planes of weakness parallel to aligned crystallites and thin, vesicle-rich layers. When the joints formed they would readily have followed such a preexisting direction of weakness. However, the presence of several sets of intersecting sheeting joints in some rocks (Fig. 25) suggests that the preconditioning of the lava by the formation of flow layers probably was convenient, but not essential to the formation of the sheeting joints, because only one set could have followed the preexisting flow layers.

Pencil joints are present where two or more closely spaced joint sets intersect to yield elongate rock fragments (Fig. 25). Pencil jointing is common at the margins of flow lobes, where en masse flowage apparently continued after the lava had devitrified.

An enigmatic type of structure that is superimposed on sheeting joints in the devitrified parts of many flows, mainly in marginal areas, is dimple joints (Fig. 26). The dimples are partially bounded by curved fractures, and individual dimples may touch at their edges, as shown in Figure 26, or be separated from one another. The dimples, which range from 5 to 40 cm in width, generally are present as groups and make complexly jointed zones.

Where sheeting joints are widely spaced or absent, the rhyolite breaks into blocky rather than platy fragments. This blocky style of breakage is most common at the very base of the central zones, around the margins of flow lobes, and in portions of the upper zones. Blocky breakage is like the fracture pattern in glassy rhyolite (Figs. 17 and 24). The locations of block-fractured zones

Figure 27. Tension fissures within a rhyolite layer from the upper zone of the Sheep Creek Rhyolite in Bruneau River canyon. These gashes, which are a few centimeters apart, developed in the wall of a cavity at the nose of a tight fold.

Figure 28. Parallel flow marks that probably developed from the collapse of elongate vesicles. Exposed on near-vertical sheeting-joint surfaces at the margin of the Marys Creek Rhyolite near Marys Creek canyon.

in devitrified rhyolite suggest that the stresses required to form sheeting joints had already dissipated and that the blocky joints had formed before devitrification occurred.

Open tension gashes a few centimeters apart (Fig. 27) are present in folded devitrified rhyolite in the basal and upper zones and marginal parts of several flows. Generally, as shown in Figure 27, these fissures are abundant in certain flow layers but are sparse or lacking in the adjacent layers. Most gashes probably formed after the rhyolite had partially or completely devitrified but while it was still malleable, because we have not observed them in folded parts of glassy rhyolite (see Figs. 13 and 18). This preferential distribution suggests that small-scale viscosity variations existed during flowage, and continued even after devitrification had started. Tension gashes are common in and near the walls of large, late-stage, gas cavities (Fig. 27).

Flow marks are sporadically present in the marginal and upper parts of some flows (Fig. 28). These streaks probably originated from the stretching and subsequent compaction of elongate vesicles. They are most common on steeply dipping sheeting joints, like the nearly vertical surface in Figure 28. They do not, however, show consistent regional primary flow-direction patterns (Bonnichsen and Citron, 1982).

Gas Cavities

Gas cavities are abundant in the upper zones of the rhyolite flows and less so in the central and basal zones. They range from vesicles less than a millimeter across to giant cavities more than a meter across. Some are spherical, others have been greatly flattened or stretched, some have flat bottoms and domal tops, and many have irregular shapes. Gas cavities of similar sizes and shapes occur together in zones that form short subhorizontal lenses and in zones that are irregular in outline or extent. The abundance of gas cavities varies markedly within single flows and from one flow to another. Some flows contain only sparse vesicles, even in their upper zones, whereas others have local zones with so many cavities, especially small ones, that they resemble scoriaceous basalt. The most vesicular rhyolite occurs at flow tops and margins and in some of the breccia fragments at the flow bases. The cavities in highly vesicular rhyolite commonly were elongated or flattened by flow. Highly stretched or flattened vesicles evidently formed before adjacent ones with equant and irregular shapes.

Giant gas cavities, some considerably more than a meter across, occur in the upper zones of the rhyolite flow interiors (Fig. 5). They typically are located in strongly deformed to even partially fragmented rhyolite. They probably formed while breakage and jostling of hot fragments occurred at the end of en masse flowage of the lava, but their lack of elongation indicates they formed after the end of forward motion. These large cavities are most abundant in the Sheep Creek (Fig. 16) and Dorsey Creek lava flows.

SEQUENCE IN WHICH PHYSICAL FEATURES FORMED

In this section we note the relative times when the physical features in the lava flows formed. Our results are summarized in Figure 29 and the features are grouped into the following process-of-formation categories: phenocryst development, flow layering, gas release, brecciation, devitrification, and joint formation. The evolutionary stages are listed in chronological order: (1) magma formation, (2) magma rise to surface, (3) eruption of lava, (4) lava flowage across surface, (5) end of flowage, and (6)

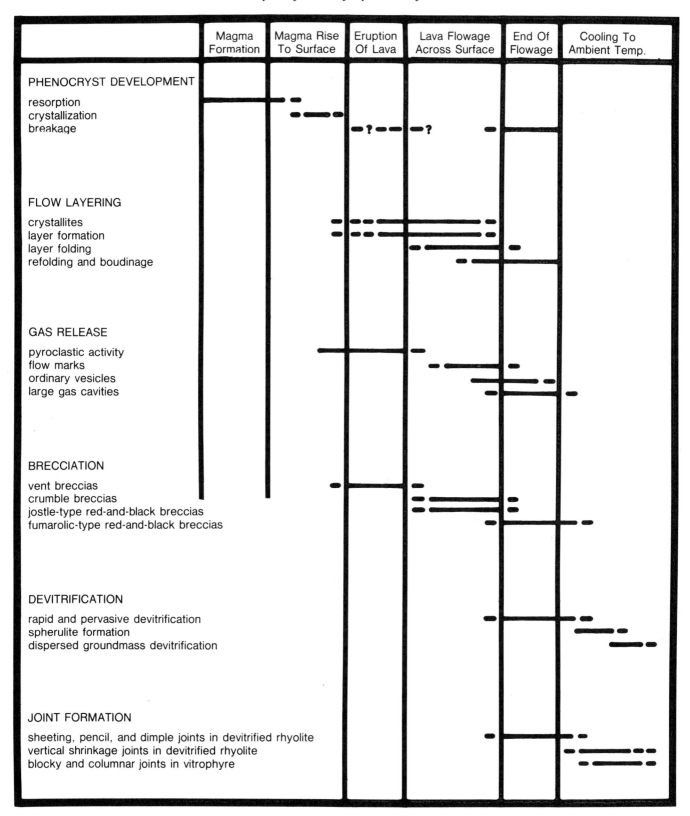

Figure 29. The approximate time sequence in which many of the features in the southwestern Idaho rhyolite lava flows formed as the magmas were generated and rose to the surface, and as the lavas were erupted, flowed across the surface, stopped, and cooled.

cooling to ambient temperatures. At a given time, different types of features may have formed concurrently in adjacent parts of the same flow.

Phenocryst Development

We believe the phenocrysts acquired many of their textural features when the magmas formed. This is especially true for plagioclase, pyroxene, and opaque oxides, which are remnant crystals and crystal fragments that were not completely resorbed when the source rocks were melted. Thus resorbtion, rather than growth, was the first major event to affect the phenocrysts. As the magmas rose, however, growth of new material on the phenocrysts may have occurred, and new phenocryst minerals—quartz and sanidine—were crystallized in some batches of magma.

Crystal breakage was the last event to modify the phenocrysts. This happened mainly in the marginal parts of the flows at the very end of flowage in conjunction with the boudinage and fracturing of flow layers. The extensively embayed plagioclase grains were especially susceptible; this is shown in a few places by pieces broken from the same phenocryst remaining near one another. Whether some plagioclase phenocrysts were broken during the eruption and flowage of the lavas is unknown, but seems probable.

Flow Layering

The flow layers in vitrophyre probably formed in response to small-scale changes in the dissolved and exsolved water contents of the silicic magmas while the pressure dropped during eruption and flowage. Adjacent areas that developed contrasting gas bubble contents and viscosities were smeared into layers during flowage. The marked variations of crystallite abundances from one flow layer to another suggest the crystallites formed concurrently with the layering, or before the layering. If the flow layers and crystallites formed as the volatiles exsolved, they probably started to form when the magmas rose near enough to the surface for vesiculation to take place and continued to form until the lavas stopped moving.

Flow layers usually are folded and refolded in glassy rocks. These features probably formed throughout en masse flowage of the lavas. The most severe disruptions—boudinage and breakage—resulted at the end of flowage from a rise in the viscosity of the glassy material; subsequent deformation must have taken the form of breakage.

Gas-Release Features

The release of volatiles from within the lava flows successively formed volcanic ash, flow marks, ordinary vesicles, and large gas cavities. The presence of small amounts of air-fall ash below and above some of the lava flows, especially near their vents, indicates enough gas was being released from some magmas as they neared the surface to cause minor pyroclastic activity. Some of this activity may have been caused by the ascending magmas intersecting and heating ground water, rather than by the release of volatiles dissolved in the magmas. Some of the water which caused the pyroclastic activity may have been meteoric, having entered the flows at their bases as they traveled over wet ground, or having been introduced from precipitation which fell onto the hot flows. The ashy matrices of the crumble and flow breccias at flow margins and bases, and of the jostle and fumarolic red-and-black breccias at the flow tops, also show that gasses were released explosively both during and after lava flowage.

The parallel flow marks probably represent gas bubbles that became greatly stretched and were collapsed before lava flowage ceased. These marks must have mainly developed before the ordinary vesicles formed because they are collapsed and the vesicles are not.

The gas cavities range from small (typically a mm to a cm across), ordinary-appearing vesicles that vary from equant to greatly elongate or flattened, to large vesicles, including giant ones a meter or more across (Fig. 16). In general, the more stretching or flattening a gas cavity has undergone, the greater its relative age. Most of the giant gas cavities are equant to irregular, rather than being flattened or elongate, implying relatively young ages. The close association of tension gashes to the large gas cavities also suggests these cavities formed late, after the lava was so viscous that it yielded partly by fracturing.

Brecciation

Several breccia types occur in the rhyolite flows (Figs. 10, 11, 12, 14, 19, 20, and 21) including some in vent areas and those interpreted as crumble breccias at flow margins and beneath flows. Two red-and-black breccia types—the jostle type with closely packed fragments and the fumarolic type with the fragments farther apart—are common in the upper zones. Brecciation of rhyolite lava took place wherever the deformation rate became too great to be accommodated by plastic flowage.

The earliest breccias were those formed during eruptions. Next were crumble breccias that formed at flow margins and basal zone breccias that represent overridden crumble breccias. The jostle type of red-and-black breccia probably formed during and at the end of flowage in response to continued motion in the flow interiors. Finally, the fumarolic type formed mainly after flowage had ceased, as indicated by the lack of plastic deformation of the breccia bodies and lack of deformation of the veins of matrix material which extend into the enclosing massive vitrophyre zones.

Devitrification

The lava flows devitrified almost completely as they cooled. Glassy rhyolite is present only where cooling was rapid at the base, top, and distal edges of flows. The temporal relationships among the physical features, field and microscopic observations of the distribution of spherulites and dispersed devitrification, and the tendency for glass to devitrify faster at higher temperatures, all suggest that most devitrification took place at quite high temperatures as the lavas were coming to rest, or shortly thereafter. This timing is also implied by the restriction of sheeting joints to devitrified rhyolite. If these subparallel joints resulted mainly from strsses that developed during the final adjustments to forward motion, then much of the devitrification must have preceded their formation.

Spherulites evidently grew in the glassy parts of the flows after the flow interiors had devitrified, although the two styles of crystal growth probably overlapped in time. The lack of deformation of the spherulites suggests that they formed after all plastic or fracture-controlled deformation of the enclosing glass had ceased. Finally, the dispersed devitrification in some glassy rocks was probably the last to develop.

Joint Formation

During and after final forward motion of the lavas, several types of joints were formed (Figs. 4, 6, 7, 8, 17, 22, 23, 24, 25, and 26), including sheeting, pencil, dimple, and shrinkage joints in devitrified rhyolite and blocky breakage and columnar joints in glassy rocks.

The sheeting joints are probably oldest because they formed in response to flow stresses. The pencil and dimple joints probably formed at the same time as sheeting joints. The pencil joints must have formed before forward motion ended, but after devitrification.

The shrinkage joints apparently formed after forward motion had ceased and after devitrification, as indicated by their lack of deformation and their absence from glassy rocks. Furthermore, cross-cutting relationships indicate that shrinkage joints formed after sheeting joints.

Blocky breakage and columnar-joint development in glassy rocks also took place as the flows cooled, probably while the shrinkage joints were forming in devitrified rhyolite. By the time the blocky and columnar joints had formed, the glassy rocks must have been too cool to deform plastically. The existence of devitrified rhyolite with the blocky breakage habit near some flow tops and lobe margins implies that pervasive devitrification must have continued even after all flowage had ceased.

COMPARISON OF RHYOLITE LAVA FLOWS WITH HIGH-TEMPERATURE ASH-FLOW SHEETS

We have described above the physical features in the southwestern Idaho rhyolite lava flows and indicated their sequence of formation. Below, we will explain how these lava flows differ from high-temperature ash flows that were emplaced in the same region.

High-Temperature Ash-Flow Sheets in Southwestern Idaho

Many southwestern Idaho rhyolite sheets have been interpreted as ash flows that were welded so densely that they resemble the rhyolite lava flows in that same region. These include the Cougar Point Tuff (Bonnichsen and Citron, 1982), and the rhyolite of Grasmere escarpment (Bonnichsen, 1982a; Kauffman and Bonnichsen, in press), as well as the tuffs of Wilson and Browns creeks along the southwestern side of the western Snake River Plain northwest of the Jacks Creek area, and the five units erupted from the Juniper Mountain volcanic center (Fig. 1) described by Ekren and others (1981, 1982, 1984).

The Cougar Point Tuff is a series of densely welded ash-flow-tuff sheets, individually up to about 100 m in thickness, exposed along the southern margin of the Bruneau-Jarbidge eruptive center (Fig. 2). Chemically and petrographically these sheets resemble the rhyolite lava flows that erupted later from the same area (Table 3). Phenocrysts in the Cougar Point Tuff are plagioclase, sanidine, quartz, pigeonite, augite, magnetite, ilmenite, and olivine. The Cougar Point Tuff units were apparently hot enough when emplaced for the ash to coalesce into liquid pools capable of flowing (Bonnichsen and Citron, 1982). Few locations exist, however, where the sheets actually flowed en masse, probably because they were deposited on flat terrain.

The Wilson Creek and Browns Creek units and the slightly older rhyolite sheets that erupted from the Juniper Mountain volcanic center (Table 3) closely resemble the rhyolite lava flows in the Bruneau-Jarbidge and Jacks Creek areas, as well as the Cougar Point Tuff, in their densely welded appearances and widespread distributions. These Juniper Mountain and western Snake River Plain units have the same phenocrysts as the rhyolite flows discussed above and the Cougar Point Tuff: plagioclase, pyroxenes, opaque oxides, quartz, and alkali feldspars, but no biotite or hornblende. Chemically, they are similar to the Cougar Point Tuff (see Table 3 and references cited therein), but are more silicic than most of the Bruneau-Jarbidge area rhyolite lava flows.

Ekren and others (1981, 1982, 1984) interpreted the rhyolite sheets erupted from the Juniper Mountain area and from the southwestern margin of the western Snake River Plain as ash flows which, after emplacement, coalesced into liquid pools and flowed en masse. They asserted that virtually all of the evidence for the sheets having flowed as ash was destroyed, but stated that the sporadic occurrences of shards and pumice, the vertical compositional zoning within some of the sheets, and their laterally widespread nature are sufficient evidence to allow *all* of these rhyolite sheets to be interpreted as ash flows. They suggest that the irregular terrain caused the sheets to undergo sufficient post-emplacement flowage for them to ultimately resemble lava flows. As we noted above, they further interpreted the Bruneau-Jarbidge and Jacks Creek area rhyolite sheets as ash flows that had flowed en masse after emplacement. We show below, however, that the Bruneau-Jarbidge and Jacks Creek area flows differ considerably in detail from welded-tuff sheets like those of the Cougar Point Tuff group and those from the Juniper Mountain area and that they do not show evidence of originating as ash flows.

Magma Conditions During Eruption and Flowage

Factors that affect the eruption and flowage styles of rhyolite lavas include the temperature of the magmas and their water

TABLE 3. AGES AND SiO_2 CONTENTS OF VARIOUS WELDED-TUFF UNITS IN SOUTHWESTERN IDAHO

Name of Unit	Area	K/A Age (Ma)	SiO_2	N	References
Cougar Point Tuff, unit XV	BJEC	---	73.7	4	2, 3, 4
Cougar Point Tuff, unit XIII	BJEC	---	75.3	2	2, 3, 4
Cougar Point Tuff, unit XII	BJEC	---	73.1	2	2, 3, 4
Cougar Point Tuff, unit XI	BJEC	---	75.3	2	2, 3, 4
Cougar Point Tuff, unit IX	BJEC	---	75.6	2	2, 3, 4
Cougar Point Tuff, unit VII	BJEC	11.3 ± 2.0	74.0	3	2, 3, 4
Cougar Point Tuff, unit V	BJEC	---	75.7	2	2, 3, 4
Cougar Point Tuff, unit III	BJEC	---	75.9	2	2, 3, 4
rhyolite, Grasmere escarpment	BJEC	11.22 ± 0.52	74.3	6	4, 6, 7, 8
tuff of Browns Creek	SWSRP	11.2 average of 2	76.6	2	1, 5, 9
tuff of Wilson Creek	SWSRP	---	76.3	1	5
tuff of The Badlands	JMVC	12.0 ± 0.2	75.4	2	1, 5, 9
upper lobes, Juniper Mountain	JMVC	13.9 ± 0.5	76.9	2	5
lower lobes, Juniper Mountain	JMVC	13.8 ± 0.5	74.7	3	5
Swisher Mountain Tuff	JMVC	13.85 average of 2	72.6	2	1, 5, 9
tuff of Mill Creek	JMVC	---	72.6	1	5

Average SiO_2 = 74.9

Notes:
1. The rhyolite units are listed in approximate order of increasing age within their respective areas. The rhyolite of Grasmere escarpment may correlate with unit VII of the Cougar Point Tuff.
2. The area abbreviations are BJEC = Bruneau-Jarbidge eruptive center, SWSRP = southwestern margin of western Snake River Plain, JMVC = Juniper Mountain volcanic center.
3. The SiO_2 values are weight % SiO_2 normalized to a sum of 100 % for the 10 major constituent oxides (SiO_2, Al_2O_3, TiO_2, Fe_2O_3, MnO, CaO, MgO, K_2O, Na_2O, and P_2O_5); all iron expressed as Fe_2O_3.
4. N is the number of analyzed rocks that were averaged for the SiO_2 values.
5. References:
   1 = Armstrong and others, 1980
   2 = Bonnichsen, 1982c
   3 = Bonnichsen and Citron, 1982
   4 = Bonnichsen, unpublished data
   5 = Ekren and others, 1984
   6 = Hart and Aronson, 1983
   7 = Kauffman and Bonnichsen, 1986
   8 = Kauffman, unpublished data
   9 = Neill, 1975

contents, which in turn affect their viscosities and effusion rates. The absence of micas and amphiboles from the southwestern Idaho rhyolite lava flows suggest that they were erupted at higher temperatures than most other rhyolites. The persistence of pigeonite, rather than hypersthene, as the Ca-poor pyroxene phenocryst mineral indicates extrusion temperatures that might have been around 1000°C (Huebner, 1980) and certainly were well above 825°C (Lindsley, 1983) for many flows. The general similarity of the rhyolite flows to the Cougar Point Tuff units suggests that the eruption temperatures for the two types of units were nearly equal. A few coexisting-oxide geothermometric determinations have been made on Cougar Point Tuff samples (Hildreth, 1981) and for a few of the rhyolite lava flows (Ekren and others, 1984). The results suggest eruption temperatures in the 950–1100°C range.

The rhyolitic magmas that gave rise to the lava flows were comparatively low in dissolved water as they approached the surface. This is indicated by their lack of hydrous minerals, the volumetrically insignificant amounts of pyroclastic products that accompany the flows, and the general lack of deuteric alteration. The persistence of glass, which was only oxidized but not hydrothermally altered or devitrified, in the matrix of the fumarolic type of red-and-black breccia also suggests the lavas were low in dissolved water.

The following observations imply that the southwestern Idaho rhyolite lavas were less viscous than those in some other rhyolitic fields and less viscous than the parent magmas of the southwestern Idaho ash-flow sheets. The rhyolite lavas of southwestern Idaho have relatively low SiO_2 contents in comparison to many other rhyolites. For example, the average SiO_2 content for the 12 flows in the Bruneau-Jarbidge eruptive center is only 71.8% (Table 1), and the average for the 10 flows listed in Table 2 is 72.6%. In comparison, the average SiO_2 content of the 16 ash-flow units listed in Table 3 is 74.9%. A reduction of about 2.5% in silica content would reduce the viscosity of a Snake River Plain volcanic province type of rhyolitic lava in the 900 to 1000°C range by about one order of magnitude when calculated for anhydrous conditions (Shaw, 1972).

The comparatively high temperatures indicated for the southwestern Idaho rhyolite lavas also imply that they had reduced viscosities. A temperature increase of about 150°C would reduce the viscosity of a rhyolitic lava by about two orders of magnitude (Bottinga and Weill, 1972; Shaw, 1972). In addition, gas bubbles in the lavas would generally tend to lower their bulk viscosities, because such gasses would be much less viscous than their enclosing melts. This would become especially important when the proportion of gas was increased to the point where the gas phase became continuous in parts of a flow (Williams and McBirney, 1979). This gas-bubble effect may have been significant in southwestern Idaho because the flows apparently gave off gasses throughout their eruption and flowage histories.

Distinctions Between Flowage of Lava and Flowage of Ash

Differences between the transport of high-temperature rhyolitic lava and ash result in differences in the rhyolite sheets that are formed. Here we review some of the general aspects of these contrasting styles of flowage.

In eruptions that result in ash flows, the gasses that exsolve from water-saturated silicic magmas as the pressure drops cause vesiculation and the breakage of the magma into bubble-wall shards, small pieces of pumice, and glassy dust, all of which erupt violently. A mass of such particles, when enclosed in a hot gas cloud, can flow for tens of kilometers from the eruption site. While traveling across the surface, such ash flows commonly pick up loose rock fragments that become incorporated within the resultant deposits. If enough heat remains after emplacement, the ash will weld and cool to a dense rock.

At even higher temperatures the ash particles will, after emplacement, coalesce into a liquid-silicate pool capable of rheomorphism, or flowing en masse (Wolff and Wright, 1981). Chapin and Lowell (1979) have dubbed this postdepositional movement as secondary flowage. They refer to flow structures that developed then as secondary, in contrast to primary structures which form during or at the end of forward flowage away from the erupting source. Once hot ash particles have coalesced into a liquid pool, the pool could flow downhill as lava under the influence of gravity. The distance it might travel would depend on the terrain, the volume of silicate liquid, its cooling rate, and its bulk viscosity.

As a mass of silicate liquid flows across the surface, most primary structures or other features that might have formed earlier during ash-flow transport would be destroyed. Particularly susceptible would be bubble-wall shards, small pumice fragments, and primary flow marks. Features that would tend to persist through secondary flowage include large pumice fragments, rock fragments picked up from the surface, especially if they were refractory enough or large enough to resist assimilation, and features frozen into glass at the base or at other flow margins. However, at extremely high temperatures even these features might become obscured.

Because of their generally high silicate-melt viscosities, rhyolite lavas normally do not travel very far. Hence, rhyolite eruptions commonly produce ash flows or volcanic domes rather than extensive lava flows. We believe that low water contents are the main reason why many of the southwestern Idaho rhyolitic magmas did not erupt violently to form ash flows, but erupted as lava. Several factors probably permitted these lavas to then flow as far as they did. These factors are eruption rate, temperature, volume, bulk viscosity, and nature of the venting source. Especially important for long-distance travel are high rates of eruption (Walker, 1973), and large volumes of lava (Malin, 1980). Larger volumes of material and faster eruption rates will increase the distance a lava can flow from its source, because the greater mass will cause a longer heat-retention period and because the taller pile of lava over the vent will push lava away with more force. Lavas erupted at comparatively high temperatures will not only have lower initial viscosities but will also have more time to flow before cooling.

The presence of gas bubbles within lava will lower its overall, or bulk, viscosity and permit it to flow more easily. Volatiles that exsolve from lava as it flows and basal gas-rich zones that form from travel over wet ground might cause subhorizontal, bubble-rich zones or even flat-lying gas layers of large lateral extent; these would be planes across which lateral motion of the flow would be enhanced. The possible existence of such gas-rich zones in the southwestern Idaho rhyolite lava flows might have been an important mechanism by which some of the travel across the surface was achieved. The common occurrences of fine-grained sedimentary materials beneath the flows, and the sporadic exposures in the flow bases where such sediments were incorporated while perhaps in a wet condition, also supports the concept that gas-rich zones may have helped the flows to travel.

Finally, the long exposures of these lava flows in the canyons of southwestern Idaho suggest that the flows must have traveled far from their sources. If the lavas were erupted from long fissures rather than from point-source vents, however, the flows would not have had to travel nearly as far as is suggested by their dimensions, especially if the lava in a single flow was erupted from more than one fissure.

Even though we have concluded that the rhyolite lava flows in the Bruneau-Jarbidge and Jacks Creek areas are not remobilized ash-flow-tuff sheets, it remains possible that, when rhyolitic magmas are erupted explosively at superheated temperatures approaching those attained in basalt flows, sufficient en masse flowage of the coalesced liquid pool might entirely destroy any evidence that the rhyolite had been emplaced as ash rather than lava. With regard to the present state of knowledge about this dilemma, it is clear that further laboratory and theoretical studies are needed to develop a better understanding of the physical processes that occur within high-temperature rhyolitic lava and ash flows, so as to arrive at new and quantitative petrologic criteria to distinguish the two flow mechanisms.

Differences Between Rhyolite Lava Flows and Ash-Flow Sheets

Because some of the rhyolite lava flows in southwestern Idaho extend for distances that are more commonly associated with ash-flow units, and because the two types of units have many similarities and can easily be confused, we have identified the following features that help distinguish the mode of emplacement for individual rhyolite sheets. In part, this discussion compares the densely welded ash-flow units of the Cougar Point Tuff with the rhyolite lava flows in the Bruneau-Jarbidge eruptive center, as considered previously by Bonnichsen and Citron (1982), and in part it compares the rhyolite lava flows with the densely welded, ash-flow tuff units in southwest Idaho that Ekren and others (1982, 1984) suggested underwent such extensive secondary flowage as to resemble lava flows.

Lithic fragments are present in the lower parts of some of the Cougar Point Tuff units; however, we have not observed any in the Bruneau-Jarbidge or Jacks Creek area rhyolite lava flows. Because of their ability to be preserved in an ash flow even if it undergoes extensive secondary flowage, lithic fragments are particularly good indicators to tell if a rhyolite sheet traveled for any substantial distance as an ash flow.

Flattened pumice fragments up to several centimeters long are abundant in places in some of the Cougar Point Tuff units (Bonnichsen and Citron, 1982) and in the lower parts of the western Snake River Plain and Juniper Mountain ash-flow units (Ekren and others, 1984), but they have not been observed in the rhyolite flows from either the Bruneau-Jarbidge or Jacks Creek areas. Swarms of somewhat elongate, parallel, black-vitrophyre breccia fragments are present in the basal zones of some of the lava flows. These superficially resemble swarms of pumice clasts. The fragments differ, however, by having blunt to angular ends, in contrast to the tapered ends of squashed pumice clasts.

The rhyolite lava flows and welded-tuff units are similar petrographically, especially where devitrified. In glassy groundmasses of ash flows, however, bubble-wall shards and small pumice fragments are easily discerned, although many have been greatly flattened and deformed. Deformed bubble-wall shards and small pumice fragments are absent from the rhyolite flows. Other types of shards, generally more blocky in shape, are present sporadically near the margins of some of the lava flows, and probably were formed where the flows interacted with water. These types of shards, however, do not occur within the glassy parts of the main bodies of the flows.

The basal vitrophyre layers in the lava flows contain numerous tightly folded flow layers, generally with subhorizontal axial planes. Such layers and folds are rare in the basal vitrophyre zones of the Cougar Point Tuff units. Ekren and others (1984) reported flow layering in the basal vitrophyres of the ash-flow units they examined.

Faulting and boudinage of flow layers are present in areas of the rhyolite lava flows where the glass viscosity evidently became very high. The resulting textures in some places resemble the deformed-particle textures found in strongly compacted welded tuffs. Glassy parts of the rhyolite lava flows also contain refolded folds which, in thin sections of certain orientations, can superficially resemble deformed shards.

Fragments of broken plagioclase and sanidine crystals are abundant in the ash flows but uncommon in the rhyolite lava flows. In the lava flows they occur mainly near the flow margins in the same rocks in which the flow layers are fractured and boudinaged. In a rhyolite sheet, the presence of abundant broken phenocrysts is strong evidence for the flowage of ash if the crystal fragments are widespread, as in the Cougar Point Tuff units, rather than being confined to the most highly strained rocks as in the lava flows.

The subhorizontal flow marks that are abundant in the Cougar Point Tuff (Bonnichsen and Citron, 1982, Figs. 19 and 20), and which were used to establish primary flow azimuths, are rare in the rhyolite lava flows. In the lava flows, the sheeting joints on which these marks are impressed typically have been rotated to steep attitudes, and the marks give no unified sense of the eruptive area location.

The distal ends of the two types of rhyolite sheets differ conspicuously. Lava flows have blunt ends and marginal flow lobes with intervening, structurally complex septa, and are typically no less than 25 m thick. Thick accumulations of crumble breccia are preserved at some flow margins. The distal parts of some of the ash-flow sheets, however, are as thin as 3 to 6 m and do not have flow lobes or breccias. None of the rhyolite lava flows has been observed to thin to only a few meters, or to bifurcate to two or more cooling units in the same fashion as some Cougar Point Tuff units (Bonnichsen and Citron, 1982).

The lava flows commonly lie directly on silty or clay-rich sedimentary materials. Volcanic-ash layers occur below a few flows, but few exceed a few centimeters in thickness and they do not persist laterally. The Cougar Point Tuff units, on the other hand, typically lie on a meter or more of thinly layered, air-fall-ash layers that have great horizontal continuity. The antidune, base-surge type of inclined layering (Crowe and Fisher, 1973) also occurs sporadically in these basal ash layers of some units, but has not been observed in the ash layers beneath the lava flows.

At the tops of basal vitrophyre zones of both the lava flows and ash-flow units, we find a similar upward transition from glassy to devitrified rhyolite that commonly includes variable amounts of spherulitic devitrification. The two types of flows differ, however, in the nature of the contact separating these glassy and devitrified zones. In the ash flows, the contact is sharp, nearly planar, and subhorizontal. In the lava flows, the contact is typically convoluted with alternating glassy and devitrified layers that are commonly folded, even though the contact as a whole may be subhorizontal.

Swarms of closely spaced lithophysae, commonly with devitrified coronas, are present in the basal vitrophyres of many Cougar Point Tuff units, but are unusual in the rhyolite lava flows. Also, the laterally persistent horizontal layers that contain abundant irregular to subequant gas cavities (lithophysal zones), some of which have been traced for many kilometers in some of the Couger Point Tuff units (Bonnichsen and Citron, 1982, Fig. 10), have not been observed in any of the rhyolite flows. The swarms of gas cavities referred to above in the rhyolite lava flows have not been observed to extend for great lateral distances, and the abundance of cavities in the lava-flow swarms is not nearly as great as within the Cougar Point Tuff lithophysal zones.

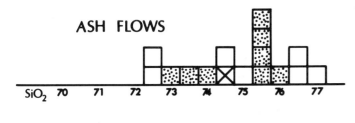

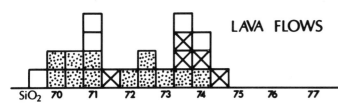

Figure 30. A comparison of the silica contents of the rhyolite lava flows and welded ash flows of southwestern Idaho. See Tables 1, 2, and 3 for data. One box per unit; stippled boxes are units erupted from the Bruneau-Jarbidge eruptive center; boxes with X are Jacks Creek area units; open boxes are other southwestern Idaho units.

Upper vitrophyre zones are common in the lava flows but not in the Cougar Point Tuff units, as are giant gas cavities and rootless dikes. The upper zones of all the lava flows have widespread red-and-black breccia, but such breccias have not been observed at the tops of any of the Cougar Point Tuff units. They are present, however, in the upper part of the rhyolite of Grasmere escarpment, a unit that probably was emplaced as ash and then flowed like lava.

In the central parts of thick rhyolite sheets, a vertical section of massive rhyolite that lacks horizontal layering arising from cooling interruptions due to a sequence of ash emplacements suggests that the entire thickness cooled as one unit. This could indicate either a lava flow or a single emplacement of ash, as might occur within a caldera. In the southwestern Idaho canyon exposures, the central zones of the rhyolite lava flows typically erode to a double cliff, separated by a medial slope but lacking other layering. In contrast, good vertical exposures of many of the Cougar Point Tuff units show horizontal layering, probably resulting from multiple ash emplacements, even though the sheets are simple cooling units.

The rhyolite lava flows are generally less siliceous than the ash flows from southwestern Idaho (Tables 1–3, Fig. 30). In addition, individual rhyolite lava flows show very little internal variation in their phenocryst mineral contents or abundances or in their chemical compositions (Bonnichsen, 1982b), whereas the ash-flow units characteristically show considerable vertical and lateral chemical variation (Ekren and others, 1984; Bonnichsen, unpublished data), and phenocryst variation (Ekren and others, 1984; Bonnichsen and Citron, 1982).

The rhyolite lava flows have breccias at their bases, especially in their marginal zones, but such breccias have not been observed in any of the Cougar Point Tuff units. The absence of a basal breccia, however, does not imply that a particular unit is an ash flow, because breccias commonly are missing under the interior parts of the rhyolite lava flows (Fig. 5). Ekren and others (1984) noted that basal flow breccias are common beneath the ash-flow units that erupted from the Juniper Mountain area. Thus, the presence of basal flow breccias only indicates that the material flowed en masse.

Sheeting joints in the lava flows and ash flows are quite similar, although in the upper zones of lava flows the sheeting joints tend to dip more steeply. The dimple and pencil joints that occur in parts of the lava flows, and which probably form as a result of en masse flowage, are virtually absent from the Cougar Point Tuff.

Diagnostic Features for Rhyolite Lava Flows

In southwestern Idaho the rhyolite lava flows can easily be confused with high-temperature, welded-tuff sheets. The features discussed above are the best indicators of whether a particular occurrence might be part of a lava flow or part of a welded tuff. In summary, the following features are useful for identifying a rhyolite sheet as a lava flow rather than a welded tuff unit. Because of the similarity in appearance of these two types of flows, several of these features should be used together.

(1) The absence of bubble-wall shards, pumice fragments, and lithic fragments, especially foreign rock types, within the rhyolite.

(2) A predominance of phenocrysts that are intact, and the presence of flow layering with evidence of its subsequent deformation.

(3) A paucity or lack of parallel flow marks that record primary flow azimuths and of laterally extensive lithophysal horizons within or at the base of the flows.

(4) Lobate marginal zones with blunt terminations and crumble-breccia accumulations.

(5) A paucity or absence of air-fall ash beneath the rhyolite sheet, except near possible vent locations, and a lack of antidune cross layering in the ash.

(6) A complexly deformed contact between the basal glassy zone and the overlying devitrified central zone of the flow and evidence, such as folds, of en masse flowage within the basal vitrophyre.

(7) The presence of extensive vitrophyre in the upper zone of the sheet, which may contain abundant areas of red-and-black breccia.

(8) A lack of internal subhorizontal layering indicative of multiple ash emplacements within the sheet, but the presence of two cliffs separated by a medial zone where the vertical shrinkage fractures from above and below are mismatched.

(9) An absence of vertical and lateral zoning in chemical composition, phenocryst types, and phenocryst abundance in the rhyolite sheet.

(10) The presence of pencil and dimple joints and breccias that suggest en masse flowage.

(11) The presence of giant gas cavities.

CONCLUSIONS

The large rhyolite lava flows in the Bruneau-Jarbidge and Jacks Creek areas of the central and western Snake River Plain volcanic province are characterized by similar physical features, phenocryst assemblages, and chemical compositions, all of which suggests they were generated and erupted under similar conditions. They have moderate to large volumes, are divisible into the same interior and marginal zones, and appear to have flowed far enough from their sources so as to show little evidence of where their eruptive fissures are located. All contain many physical features indicative of flowage across the surface as lava, and virtually no evidence exists in any which suggests they could be remobilized ash-flow sheets.

Features in the lava flows that appear to have formed prior to their eruption include phenocrysts, and possibly some of the crystallites. The phenocrysts represent both material brought to the surface from where the magma formed and crystals formed during the rise of the magma. Features that formed as the rhyolitic lavas erupted and flowed across the surface include flow layering, crystallites, breccias, and gas cavities. As the flows stopped and cooled, most of their characteristic features formed, including devitrification, sheeting, pencil, and dimple joints, various sizes of vesicles, columnar and shrinkage joints, and additional breccia deposits. The flows cooled to ambient temperatures rapidly enough so that glassy material is present at the tops, bases, and margins of the flows.

It appears to us that a combination of large magma volumes, high effusion rates, high eruption temperatures, and low initial water contents in the southwestern Idaho rhyolitic magmas imparted low enough bulk viscosities to the lavas for them to flow away from their eruptive fissures, rather than, as in other rhyolitic provinces, to pond over the eruptive fissures as steep-sided domes.

ACKNOWLEDGMENTS

We thank Dale Conover, Falma Moye, Carol Bonnichsen, Everett Bonnichsen, and James Bonnichsen for their assistance during some parts of the field work. We also thank Bill Leeman, Roger Stewart, and Jon Fink for their helpful reviews of our manuscript, and Melinda Nichols and Vicki Mitchell for their assistance with manuscript preparation. We are grateful to Margaret Jenks for her help in the field, with drafting, and in reviewing and editing this and a previous version of our manuscript.

REFERENCES CITED

Armstrong, R. L., Harakal, J. E., and Neill, W. M., 1980, K-Ar dating of Snake River Plain (Idaho) volcanic rocks; new results: Isochron/West, no. 27, p. 5–10.

Bernt, J., and Bonnichsen, B., 1982, Pre-Cougar Point Tuff volcanic rocks near the Idaho-Nevada border, Owyhee County, Idaho, in Bonnichsen, B., and Breckenridge, R. M., eds., Cenozoic geology of Idaho: Idaho Bureau of Mines and Geology Bulletin 26, p. 321–330.

Bonnichsen, B., 1981, Stratigraphy and measurements of magnetic polarity for volcanic units in the Bruneau-Jarbidge eruptive center, Owyhee County, Idaho: Idaho Bureau of Mines and Geology Technical Report 81-5, 75 p.

—— , 1982a, The Bruneau-Jarbidge eruptive center, southwestern Idaho, in Bonnichsen, B., and Breckenridge, R. M., eds., Cenozoic geology of Idaho: Idaho Bureau of Mines and Geology Bulletin 26, p. 237–254.

—— , 1982b, Rhyolite lava flows in the Bruneau-Jarbidge eruptive center, southwestern Idaho, in Bonnichsen, B., and Breckenridge, R. M., eds., Cenozoic geology of Idaho: Idaho Bureau of Mines and Geology Bulletin 26, p. 283–320.

—— , 1982c, Chemical compositions of the Cougar Point Tuff and rhyolite lava flows from the Bruneau-Jarbidge eruptive center, Owyhee County, Idaho: Idaho Bureau of Mines and Geology Technical Report 82-1, 22 p.

—— , 1985, Neogene rhyolitic volcanism, southwestern Idaho and vicinity: Geological Society of America Abstracts with Programs, v. 17, no. 4, p. 210.

Bonnichsen, B., and Citron, G. P., 1982, The Cougar Point Tuff, southwestern Idaho and vicinity, in Bonnichsen, B., and Breckenridge, R. M., eds., Cenozoic geology of Idaho: Idaho Bureau of Mines and Geology Bulletin 26, p. 255–281.

Bonnichsen, B., and Jenks, M. D., 1986, Geologic map of the Jarbidge River Wilderness Study Area, Owyhee County, Idaho: U.S. Geological Survey Miscellaneous Field Investigations Map, scale 1:50,000 (in press).

Bonnichsen, B., Bernt, J., and Jenks, M. D., 1986, Geologic map of the Sheep Creek West Wilderness Study Area, Owyhee County, Idaho: U.S. Geological Survey Miscellaneous Field Investigations Map, scale 1:50,000 (in press).

Bottinga, Y., and Weill, D. F., 1972, The viscosity of magmatic silicate liquids; a model for calculation: American Journal of Science, v. 272, p. 438–475.

Chapin, C. E., and Lowell, G. R., 1979, Primary and secondary flow structures in ash-flow tuffs of the Gribbles Run paleovalley, central Colorado, in Chapin, C. E., and Elston, W. E., eds., Ash-flow tuffs: Geological Society of America Special Paper 180, p. 137–154.

Christiansen, R. L., 1982, Late Cenozoic volcanism of the Island Park area, eastern Idaho, in Bonnichsen, B., and Breckenridge, R. M., eds., Cenozoic geology of Idaho: Idaho Bureau of Mines and Geology Bulletin 26, p. 345–368.

Citron, G. P., 1976, Idavada ash flows in the Three Creek area, southwestern Idaho, and their regional significance [M.S. thesis]: Ithaca, New York, Cornell University, 83 p.

Coats, R. R., 1968, The Circle Creek rhyolite, a volcanic complex in northern Elko County, Nevada, in Coats, R. R., Hay, R. L., and Anderson, C. A., eds., Studies in volcanology (Williams volume): Geological Society of America Memoir 116, p. 69–106.

Crowe, B. M., and Fisher, R. V., 1973, Sedimentary structures in base-surge deposits with special reference to cross bedding, Ubehebe Craters, Death Valley, California: Geological Society of America Bulletin, v. 84, p. 663–682.

Dalrymple, G. B., 1979, Critical tables for conversion of K-Ar ages from old to new constants: Geology, v. 7, p. 558–560.

Ekren, E. B., McIntyre, D. H., Bennett, E. H., and Malde, H. E., 1981, Geologic map of Owyhee County, Idaho, west of 116° west longitude: U.S. Geological Survey Miscellaneous Investigations Series Map I-1256, scale 1:125,000.

Ekren, E. B., McIntyre, D. H., Bennett, E. H., and Marvin, R. F., 1982, Cenozoic stratigraphy of western Owyhee County, Idaho, in Bonnichsen, B., and Breckenridge, R. M., eds., Cenozoic geology of Idaho: Idaho Bureau of Mines and Geology Bulletin 26, p. 215–235.

Ekren, E. B., McIntyre, D. H., and Bennett, E. H., 1984, High-temperature, large-volume, lavalike ash-flow tuffs without calderas in southwestern Idaho: U.S. Geological Survey Professional Paper 1272, 76 p.

Hart, W. K., and Aronson, J. L., 1983, K-Ar ages of rhyolite from the western

Snake River Plain area, Oregon, Idaho, and Nevada: Isochron/West, no. 36, p. 17–19.

Hildreth, W., 1981, Gradients in silicic magma chambers; implications for lithospheric magmatism: Journal of Geophysical Research, v. 86, p. 10153–10192.

Huebner, J. S., 1980, Pyroxene phase equilibria at low pressure, *in* Prewitt, C. T., ed., Pyroxenes: Mineralogical Society of America Reviews in Mineralogy, v. 7, p. 213–288.

Jenks, M. D., and Bonnichsen, B., 1986, Geologic map of the Bureau River Wilderness Study Area, Owyhee County, Idaho: U.S. Geological Survey Miscellaneous Field Investigations Map, scale 1:50,000 (in press).

Kauffman, D. F., and Bonnichsen, B., 1986, Geologic map of the Little Jacks Creek, Big Jacks Creek, and Duncan Creek Wilderness Study Areas, Owyhee County, Idaho: U.S. Geological Survey Miscellaneous Field Investigations Map, scale 1:50,000 (in press).

Kittleman, L. R., Green, A. R., Hagood, A. R., Johnson, A. M., McMurray, J. M., Russell, R. G., and Weeden, D. A., 1965, Cenozoic stratigraphy of the Owyhee region, southeastern Oregon: University of Oregon Museum of Natural History Bulletin 1, 45 p.

Leeman, W. P., 1982, Geology of the Magic Reservoir area, Snake River Plain, Idaho, *in* Bonnichsen, B., and Breckenridge, R. M., eds., Cenozoic geology of Idaho: Idaho Bureau of Mines and Geology Bulletin 26, p. 369–376.

Lindsley, D. H., 1983, Pyroxene thermometry: American Mineralogist, v. 68, p. 477–493.

Malin, M. C., 1980, Length of Hawaiian lava flows: Geology, v. 8, p. 306–308.

Neill, W. M., 1975, Geology of the southeastern Owyhee Mountains and environs, Owyhee County, Idaho [M.S. thesis]: Stanford, California, Stanford University, 59 p.

Shaw, H. R., 1972, Viscosities of magmatic silicate liquids; an emperical method of prediction: American Journal of Science, v. 272, p. 870–893.

Spear, D. B., and King, J. S., 1982, Geology of Big Southern Butte, Idaho, *in* Bonnichsen, B., and Breckenridge, R. M., eds., Cenozoic geology of Idaho: Idaho Bureau of Mines and Geology Bulletin 26, p. 395–403.

Struhsacker, D. W., Jewell, P. W., Zeisloft, J., and Evans, S. H., Jr., 1982, The geology and geothermal setting of the Magic Reservoir area, Blaine and Camas counties, Idaho, *in* Bonnichsen, B., and Breckenridge, R. M., eds., Cenozoic geology of Idaho: Idaho Bureau of Mines and Geology Bulletin 26, p. 377–393.

Walker, G.P.L., 1973, Length of lava flows: Royal Society of London Philosophical Transactions, ser. A, v. 274, p. 107–118.

Williams, H., and McBirney, A. R., 1979, Volcanology: San Francisco, Freeman, Cooper and Co., 397 p.

Wolff, J. A., and Wright, J. V., 1981, Rheomorphism of welded tuffs: Journal of Volcanology and Geothermal Research, v. 10, p. 13–34.

MANUSCRIPT ACCEPTED BY THE SOCIETY AUGUST 26, 1986

Typeset by WESType Publishing Services, Inc., Boulder, Colorado
Printed in U.S.A. by Malloy Lithographing, Inc., Ann Arbor, Michigan